Systems Engineering

Foundations and Frameworks Series

The Engineering Framework Behind Society's Most Complex Technologies: Designing, Managing & Optimizing Complex Systems

Martin Foster

TME Press email: support@tme.net

Introduction

On September 23, 1999, NASA lost a spacecraft because two teams used different units of measurement. The Mars Climate Orbiter had traveled about 416 million miles over nine and a half months. As it approached the planet to settle into orbit, the navigation software expected engine data in metric units, newton-seconds. The software that produced that data, built by a different contractor, reported it in imperial units, pound-force seconds. Nobody caught the mismatch. The orbiter came in around 57 kilometers above the Martian surface instead of the planned 226 kilometers, far too low to survive. It burned up in the atmosphere or skipped off into space. The price of that one inconsistency was roughly $327 million.

Otherwise, every individual piece of that mission worked. The engines were sound. The software ran without crashing. The trajectory math was correct on each team's own terms. The failure lived in the space between the pieces, in the handoff where one group's assumptions didn't match another's. No single engineer made an obvious mistake. The whole thing failed even though its parts performed as designed.

That gap, the place where well-built components meet and sometimes break, is what systems engineering exists to handle. As products and infrastructure have grown more tangled, the hardest problems have moved out of the individual parts and into the connections between them. A modern car contains around 100 million lines of code spread across dozens of computers from different suppliers. A commercial aircraft has roughly six million parts. A power grid links thousands of generators, substations, and control centers that have to stay in balance every second of every day. In systems like these, you can verify each component on its own and still end up with something that fails in service, because the behavior that matters only shows up when everything runs together.

Systems engineering is the discipline that treats the whole as the thing to be designed, not just the parts. It asks a set of stubborn questions that are easy to skip when you are deep in one corner of a project. What does this system actually need to do? Who decides? How will the pieces fit together, and who owns each seam? How do we know it works before millions of dollars or human lives depend on it? How do we make decisions about cost, schedule, and performance when improving one almost always damages another?

These questions sound simple. Answering them well, on a project with hundreds of people and years of work, is genuinely hard. That difficulty is the subject of this book.

This book is written for people who build, manage, or study complex systems and want a clear map of how the discipline fits together. You don't need an advanced degree to follow it. If you're an engineer moving into a systems role, a project manager trying to understand what your technical leads are worried about, a student, or someone curious about how large technologies get made without falling

Outline

Contents

Chapter 1: Foundations of Systems Engineering

The phrase "systems engineering" first showed up at Bell Telephone Laboratories in the 1940s, and the reason it appeared there is worth understanding. By that point the American telephone network had grown into something no single person could hold in their head. Millions of phones connected through switching offices, long-distance lines, repeaters, and signaling equipment, all built by different groups over different decades. An engineer could design a flawless switch, and it would still cause trouble once it joined the larger network, because the network behaved in ways that none of its individual parts predicted. Bell Labs needed people whose job wasn't any single component but the behavior of the whole. They gave that job a name, and the discipline grew from there.

That origin tells you what systems engineering is really about. It came into being not because engineering got better, but because the things engineers built got more connected, and the trouble started living in the connections. This chapter lays out the ideas the rest of the book depends on. You'll learn what the discipline actually covers, why thinking in systems differs from thinking in straight lines, how a system moves through its life from first sketch to final shutdown, why the people who want the system rarely agree with each other, how systems produce behavior no one designed, and where a system stops and its surroundings begin. None of these ideas is complicated on its own. The work, and the difficulty, comes from holding all of them at once.

Definition and Scope of Systems Engineering

A good way into the definition is to notice what systems engineering is not. It isn't a specialty like electrical engineering or mechanical engineering, where you go deep into one kind of physics. It sits on top of those specialties and connects them.

A mechanical engineer asks whether a bracket will hold. An electrical engineer asks whether a circuit will carry the load. A systems engineer asks a different sort of question: given everything this product has to do, and everyone who has a stake in it, how do all the parts fit together so the finished thing works in the real world, on time, and within budget?

The International Council on Systems Engineering, the professional body most practitioners belong to, describes the field as a way to design, build, and manage systems over their whole life, drawing on many engineering and management disciplines at once. That's accurate but abstract, so let's make it concrete. **Systems engineering** is the practice of treating a system as a single thing to be designed, even though that thing is made of many parts built by many people. It covers the full span of work: figuring out what the system needs to do, deciding how to break it into pieces, defining how those pieces talk to each other, confirming that the assembled result actually works, and steering the cost, schedule, and technical trade-offs along the way.

The scope is wider than most people expect when they first hear the term. It isn't limited to spacecraft and missiles, even though aerospace and defense did the most to develop the early methods. Hospitals run on systems that link patient records, lab equipment, scheduling, billing, and staff, and someone has to make those pieces work together or patients get hurt. A modern car is a rolling system of dozens of computers, sensors, and actuators from a dozen suppliers, running software that has to behave consistently whether the road is icy or the battery is low. Power grids, water treatment plants, air traffic control, banking networks, and the supply chains that put groceries on shelves are all systems in this sense, and all of them employ people who do systems engineering whether or not that's the title on their badge.

To see the scope in one product, look at the **Boeing 787 Dreamliner**. Boeing decided to build it using a global supply chain, with major sections of the aircraft designed and built by partner companies on three continents. The wings came from Japan, parts of the fuselage from Italy and the United States, the landing gear from France and Britain. The idea was that each partner would deliver a finished, tested section, and Boeing would join them together quickly at the end.

The engineering of each section was sound. What went wrong was the systems-level work: the interfaces between sections didn't always match, partners fell behind without Boeing seeing it early enough, and parts arrived that didn't fit together as planned. Some sections showed up with temporary fasteners holding them together, work that was supposed to be finished. The program ran about three years late and cost billions more than projected. Every supplier could point to good engineering on their piece. The failures lived between the pieces, and managing between the pieces is exactly what systems engineering exists to do.

This points to a defining feature of the discipline. Systems engineering owns the seams. When a problem can be traced to a single component, a specialist engineer usually handles it. When a problem comes from how components interact, from a requirement nobody pinned down, from an interface two teams understood differently, or from a behavior that only appears once everything runs together, that's systems engineering territory. The harder and more connected a project gets, the more of its trouble concentrates in those seams, which is why the field keeps growing as products grow more complex.

There's a management side to the work that's just as real as the technical side. A systems engineer spends a lot of time deciding what matters most, where to spend limited money and time, which requirements to hold firm and which to relax, and how to keep hundreds of people pointed at the same target. Making a system lighter usually makes it more expensive. Making it more reliable usually makes it slower to build. Adding a feature the customer loves might break a constraint the regulator imposed. These tensions don't have clean answers, and resolving them sensibly, again and again, across the life of a program, is a large part of the job. The technical knowledge tells you what's possible. The judgment tells you what to actually do.

It helps to fix the difference between a systems engineer and the specialists they work with, because the roles get confused.

Aspect	**Specialty engineer (electrical, mechanical, software, etc.)**	**Systems engineer**
Main question	Will my component meet its spec?	Will the whole system meet the need?
Focus	Depth in one domain	Connections across all domains
Owns	A part or subsystem	Interfaces, requirements, and overall behavior
Optimizes for	The best version of that part	The best balance for the system
Worries about	Component failure	Failures that emerge between parts
Typical tools	Domain analysis, detailed design	Requirements, trade studies, interface control

The two roles need each other. A systems engineer who tries to do the specialists' deep work will get it wrong, and specialists who ignore the systems view will build excellent parts that don't add up to a working whole. The discipline succeeds when these people respect what the others see that they can't.

Systems Thinking Versus Linear Thinking

If you take away one habit of mind from this book, make it this one. **Systems thinking** is the ability to see a problem as a web of parts that affect each other, often in loops, rather than as a straight chain of cause and effect. It's the opposite of how most of us are trained to reason, and the gap between the two ways of thinking explains a surprising share of expensive failures.

Linear thinking works like this: A causes B, B causes C, so if you want more C, make more A. It's how we handle most everyday problems, and for simple situations it's correct. If your room is cold, turn up the heat. If sales are low, advertise more. The logic runs in one direction, and you can trace it cleanly from start to finish. The trouble is that complex systems almost never behave this way, and applying linear logic to them produces confident decisions that backfire.

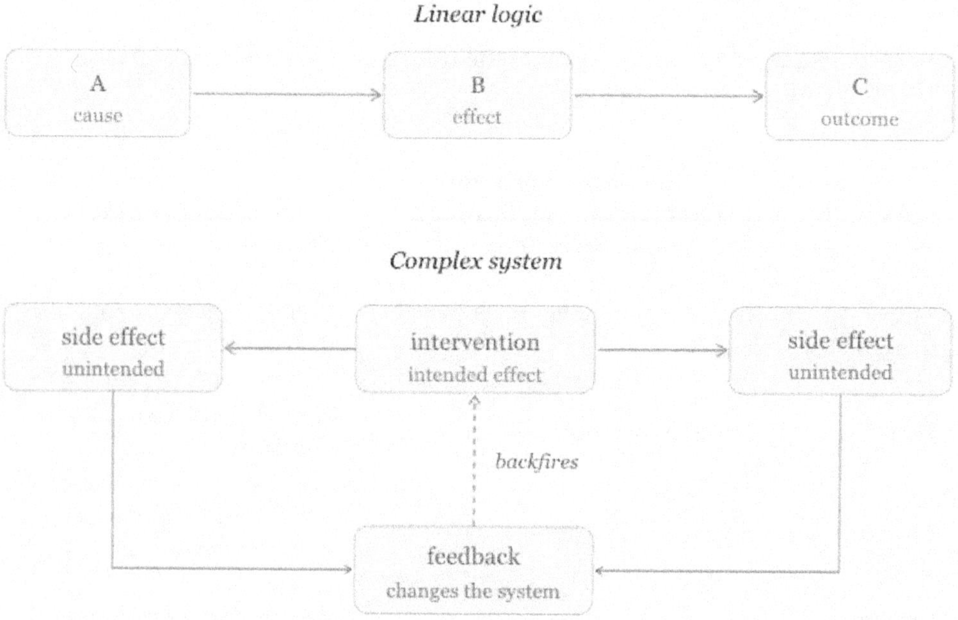

More A does not reliably produce more C

Consider traffic, which everyone has an intuition about and everyone gets wrong. The linear thought is obvious: roads are congested, so build more roads, and congestion will drop. Cities have spent enormous sums on this logic. What actually happens is well documented and has a name, **induced demand**. New road capacity lowers the cost of driving, so more people drive, so the new capacity fills up, often within a few years, leaving congestion about where it started but with more cars and more sprawl. There's an even sharper version called **Braess's paradox**, described by the mathematician Dietrich Braess in 1968, in which adding a new road to a network can make everyone's travel time worse, not better, because the new route changes how drivers choose paths in a way that clogs the whole system. Cities including Seoul and New York have seen traffic improve after closing roads, which makes no sense at all under linear thinking and perfect sense once you see the network as a system that responds to changes.

The piece that linear thinking misses is the **feedback loop**, the situation where an effect circles back and changes its own cause. A thermostat is the simple textbook case: the heater warms the room, the warmth trips the sensor, the sensor shuts off the heater, and the loop keeps temperature steady. That's a balancing loop, one that pushes toward stability. The dangerous loops are reinforcing ones, where an effect amplifies itself. In a panic, falling stock prices scare investors into selling, which drives prices down further, which scares more selling. Each turn of the loop makes the next turn stronger, and the result can be a crash far larger than any single piece of bad news would justify.

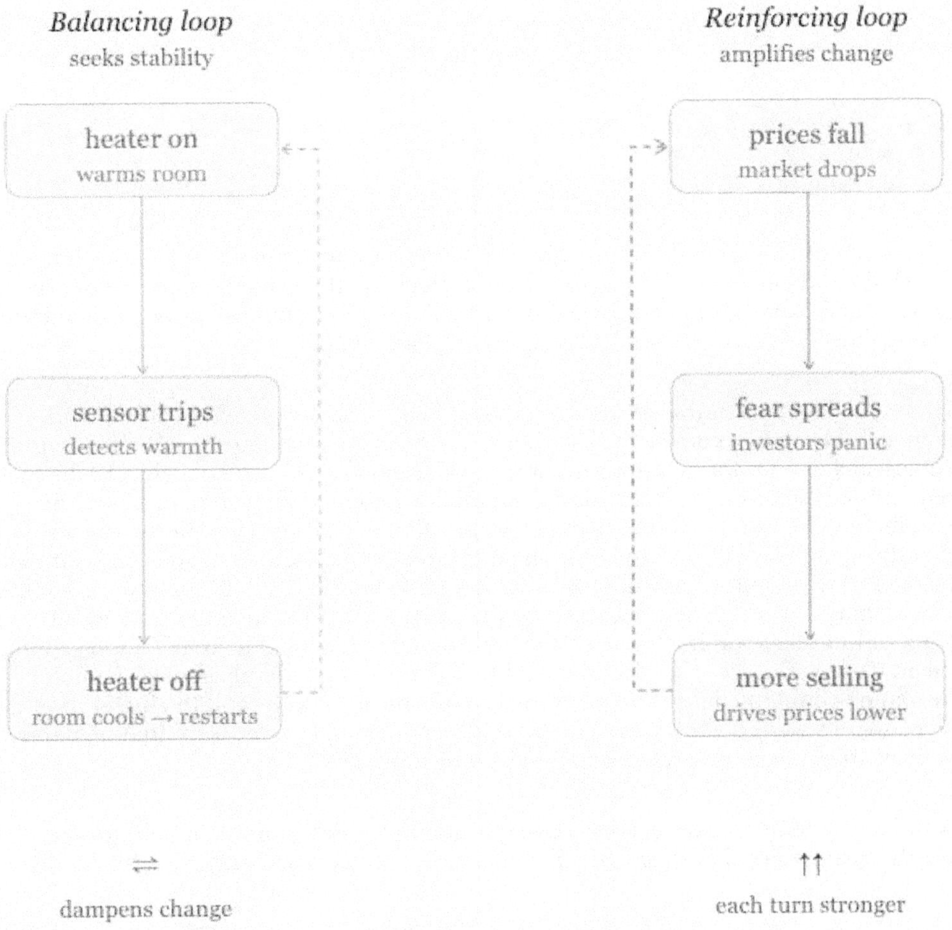

Supply chains show a feedback effect so reliable it has its own name, the **bullwhip effect**. Picture a retailer who sees a small bump in customer demand and orders a bit extra to be safe. The distributor sees the retailer's larger order, assumes demand is rising, and orders even more from the factory to build a buffer. The factory sees a big jump and ramps up production hard. A tiny wiggle at the customer end becomes a violent swing at the factory end, with everyone overreacting to everyone else's overreaction. No one in the chain is being foolish. Each link makes a reasonable local decision. The wild swing is a property of the system, not of any one decision-maker, and you can only see it if you look at the whole chain at once.

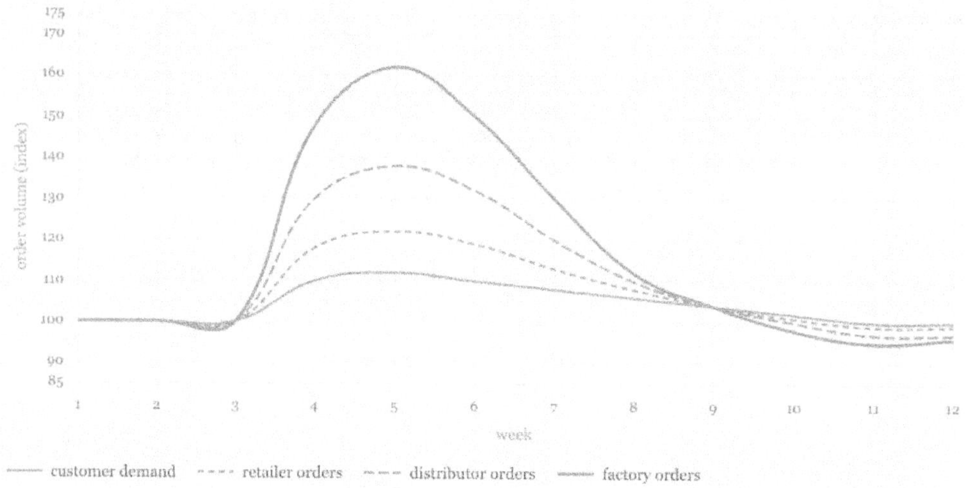

(The same small demand bump – about a 10% rise starting in week 4 – travels up the chain and becomes a 62% spike at the factory. By the time the signal settles back down, the factory has already overshot badly in both directions. Customer demand barely moves; the whip crack happens entirely upstream.)

Here's the failure that makes the lesson stick. When the U.S. federal government launched **Healthcare.gov** on October 1, 2013, the website meant to let millions of Americans sign up for health insurance, it collapsed almost immediately. On the first day, out of millions of visitors, only a handful of people managed to complete an enrollment. The site stayed broken for weeks. The official reviews afterward found that the cause wasn't one bad component. Dozens of contractors had each built their piece, and many pieces worked fine in isolation. What no one had done was treat the whole thing as a single system, test how the parts performed together under real load, or put one team in charge of the end-to-end behavior. The registration system, the identity-verification service, the data hub connecting federal agencies, and the insurance-company links were each somebody's responsibility, and the integration of all of them was nobody's. That's linear thinking applied to a problem that demanded systems thinking, and it failed in public at enormous cost.

A trained systems thinker carries a few questions into every problem, and they're worth learning because they catch the errors that linear reasoning produces.

- Where are the loops? What effects circle back to change their own causes, and are those loops pushing toward stability or runaway?
- What's the delay? Causes and effects in systems are often separated in time, so the result of a decision may not appear for months, long after you've concluded it didn't work and tried something else.
- What will the parts do in response? People and components adapt to changes, so a fix that assumes everyone holds still often fails once everyone reacts.
- Am I solving the symptom or the cause? Pushing hard on a symptom frequently relocates the problem rather than removing it, and the relocated problem is often worse and harder to trace.

That last question deserves a moment, because it names one of the most common and costly mistakes in engineering and management both. Systems tend to fight back against pressure on a symptom. Add capacity to relieve a bottleneck, and the bottleneck often moves downstream to the next-weakest point. Speed up one stage of a process, and work piles up at the next stage that can't keep pace. The system finds a new way to express the same underlying limit. Real fixes usually require finding the leverage point, the place where a modest change shifts the whole system's behavior, and those points are frequently not where the symptom shows up. Finding them is hard, which is why so many organizations keep treating symptoms and keep being surprised when the trouble returns somewhere new.

None of this means linear thinking is useless. For simple, bounded problems it's faster and perfectly correct, and a systems thinker who refuses to ever reason in a straight line is just as broken as a linear thinker who can't see a loop. The skill is knowing which situation you're in. When the parts of a problem affect each other, when there are loops and delays and adaptation, when the system will respond to what you do, you need systems thinking. The cost of using the wrong mode isn't a small error. As Healthcare.gov showed, it can be total failure of something that every individual contributor built correctly.

The System Lifecycle Model

Every system has a life, and that life follows a shape that's similar enough across domains that it's worth learning as a single pattern. The **system lifecycle** is the full span from the first idea that a system might be needed, through its design and construction, into the years it spends doing its job, and finally to the day it's shut down and disposed of. Systems engineers think in lifecycle terms constantly, because decisions made early in the life of a system lock in costs and constraints that everyone lives with for decades.

The life of a system breaks into stages that flow roughly in order, though real projects loop back and overlap more than any neat diagram suggests.

It starts with **concept**, the stage where you decide whether the system is worth building at all and what it would broadly need to do. This is where stakeholders argue about the problem, rough requirements take shape, and a few candidate approaches get sketched and compared. Nothing is built yet, and that's the point. Decisions here are cheap to change and expensive to get wrong, because a flawed concept carried forward gets harder and costlier to fix at every later stage.

Next comes **development**, where the concept becomes a detailed design and then physical or coded reality. Requirements get pinned down precisely, the system is broken into subsystems and components, interfaces between them are defined, and engineers build and test the pieces. This is the longest and most expensive stage for most systems, and it's where the bulk of the engineering labor lives.

Then **production**, where the system is manufactured, assembled, or installed at the scale it'll be used. For a one-off like a bridge or a spacecraft, this means building the single article. For a car or a phone, it means setting up a manufacturing line that turns out millions of identical units, which is its own deep engineering problem.

After that comes the long middle of a system's life, **operation and support**, the years or decades when the system does the work it was built for. This stage usually lasts far longer than all the others combined, and it's where most of the total cost actually lands, in maintenance, repairs, upgrades, fuel, staffing, and the steady work of keeping the thing running. A combat aircraft might take a decade to develop and then serve for forty years. A power plant might run for half a century.

Finally **retirement**, the stage where the system is taken out of service and disposed of. People often forget this stage exists until it arrives, and then discover it's harder than they thought. Decommissioning a nuclear plant takes decades and costs billions. Scrapping a chemical facility means handling toxic materials safely. Retiring a software system means migrating data nobody fully understands anymore. A well-designed system plans for its own end from the beginning.

The single most important fact about the lifecycle, and the reason systems engineers care about it so much, is how cost and freedom shift as you move through the stages. Early on, you have enormous freedom to shape the system and very little money committed. Late in development and beyond, you've spent a great deal but have almost no freedom left, because changing anything means undoing work that's already done.

The numbers attached to this are striking and consistent across industries. By the end of the concept stage, when only a small fraction of the total budget has been spent, the decisions made already commit something like 70 to 85 percent of the system's entire lifecycle cost. The choices that determine what a system will cost to build, run, and eventually retire are mostly locked in while it still looks like nothing but drawings and arguments.

This is why finding a problem late is so brutal. A requirements mistake caught during the concept stage costs almost nothing to fix; you change a sentence. The same mistake caught during development costs more, because you have to redo design work. Caught during production, it's worse still, because tooling and parts are involved. Caught after the system is in service, the cost can be hundreds or thousands of times the cost of fixing it at the start, because now you're recalling products, retrofitting fielded units, or living with the flaw. Studies in software and hardware both put the multiplier in this range, and while the exact figures vary, the shape of the curve never does. Errors get exponentially more expensive the longer they survive.

Lifecycle stage	Budget spent so far	Freedom to change	Cost to fix an error
Concept	Very low	Very high	Lowest
Development	Moderate	Moderate, falling	Moderate
Production	High	Low	High
Operation & support	Highest (cumulative)	Very low	Very high
Retirement	Final costs	None	N/A

The practical lesson the lifecycle teaches is to push hard, intellectually, at the front. Spend real effort getting the concept and the requirements right while changes are still cheap, even though it feels slow and abstract and there's pressure to start building something visible. The teams that rush past the early stages to show quick progress almost always pay for it later, at a far higher exchange rate. The lifecycle view is what gives a systems engineer the spine to insist on careful early work against the natural urge to just start building.

Stakeholders and Their Competing Requirements

A system exists to satisfy people, and those people are called **stakeholders**: anyone who has a stake in the system, meaning anyone who affects it or is affected by it. The reason this gets a section of its own is that beginners tend to imagine a single, sensible customer who knows what they want. Real systems serve a crowd of stakeholders who want different things, don't always know what they want, and frequently want things that flatly contradict each other. Sorting out that mess is a large part of why systems engineering is hard.

Take a single passenger aircraft and count the stakeholders. The airline that buys it wants low purchase price, low fuel burn, low maintenance cost, and high seat count. Passengers want comfort, legroom, quiet, and a low ticket price, some of which fight the airline's wishes directly. Pilots want the plane to be easy and forgiving to fly. Mechanics want it easy to inspect and repair, with parts they can reach without removing half the aircraft. Regulators like the FAA want it provably safe, which adds weight and cost. The airports it'll use want it to fit their gates and not be too loud for the neighbors. The manufacturer wants it profitable to build and possible to produce on schedule. Every one of these groups is reasonable, and almost none of them can be fully satisfied without disappointing another. More legroom means fewer seats means less airline revenue means a less attractive product. Lighter weight for fuel economy can conflict with the structural margin that makes mechanics and regulators comfortable. The systems engineer's job isn't to please everyone, which is impossible, but to find the balance that serves the whole adequately and to make the trade-offs deliberately rather than by accident.

Watch what happens when stakeholder needs aren't reconciled, in one of the most famous projects ever built. The **Sydney Opera House** was first estimated at about seven million Australian dollars, with a planned opening in 1963. It opened in 1973, ten years late, at a final cost of roughly 102 million dollars, around fourteen times the original budget. Part of the trouble was that the architect's stunning shell design was approved before anyone knew how to actually build it, so engineering and construction were chasing a form that hadn't been worked out.

But a large part of the trouble was stakeholder conflict that was never resolved up front. The government wanted it cheap and fast. The architect, Jørn Utzon, wanted his vision built without compromise. The performing groups who'd use the building had requirements about the halls that clashed with the design. These tensions were left to fight it out during construction instead of being settled during planning, and the result was redesigns, delays, a public feud, and Utzon resigning before the building was finished. The Opera House became a beloved landmark, but it's also a standing lesson in what unmanaged stakeholder conflict costs.

The work of dealing with stakeholders has a clear shape, even if the doing is messy. First you have to find them all, which sounds easy and isn't, because the stakeholders who get forgotten are the ones who cause the worst surprises later. The mechanic who can't reach a part, the neighbor who complains about noise, the regulator whose rule you didn't know about: these are stakeholders whose needs become expensive problems precisely because nobody thought to ask them early. Then you have to draw out what each one actually needs, which is harder still, because stakeholders often can't articulate their real requirements and will tell you what they think they want instead of what would actually solve their problem. A stakeholder who asks for a faster horse, in the old line, really needs to get somewhere quickly, and a good systems engineer digs past the stated request to the underlying need.

Then comes the part that has no clean solution: reconciling the conflicts. When two stakeholders want incompatible things, somebody has to decide whose need wins, by how much, and why. This is where the discipline gets political as much as technical, because the answer isn't found by calculation. It's negotiated. The systems engineer's contribution is to make the conflicts visible, lay out what each option costs whom, and force the trade-off into the open where it can be decided on purpose by people with the authority to decide it. The alternative, which is what sank parts of the Opera House and many lesser projects, is to leave the conflicts buried and let them surface as crises during construction, when fixing them is far more painful.

There's a category of stakeholder that's especially easy to shortchange: the ones who can't speak up during design because they don't exist yet or aren't in the room. Future maintainers who'll inherit the system in twenty years. The public who'll live near it. The environment that'll absorb its waste. The operators who'll work night shifts running it. These quiet stakeholders rarely have a representative at the planning table, and their needs get neglected unless someone deliberately stands in for them. Part of doing systems engineering well is remembering the people whose interests no one is paid to defend, because the consequences of ignoring them are real even when their voices aren't loud.

Emergence and System Behavior

Now we reach the idea that makes systems genuinely strange, and genuinely interesting. **Emergence** is the appearance of behavior at the level of the whole system that none of the parts has on its own and that you can't predict by studying the parts in isolation. The whole does something the pieces never could, and the something isn't always good.

The clean way to grasp emergence is that the system is more than the sum of its parts, and "more" can mean better or worse. Water is the easy positive example: hydrogen burns, oxygen feeds fire, and yet two hydrogen atoms bonded to one oxygen atom put fires out. Wetness isn't a property of hydrogen or of oxygen. It emerges only when they combine in that arrangement. You could study a single hydrogen atom forever and never find a hint of wetness in it. The behavior exists at the level of the molecule, not the atom, and that's emergence in its simplest form.

Engineered systems produce emergence constantly, and the dangerous kind is what keeps systems engineers up at night. The textbook disaster is the **Tacoma Narrows Bridge**, which opened in Washington State in July 1940 and tore itself apart in a windstorm four months later, on November 7. Every component of that bridge was sound. The steel was good, the cables were strong, the deck was properly built. But the bridge as a whole had a behavior nobody designed and nobody anticipated: in a steady moderate wind, around 40 miles per hour, the deck began to twist, and the twisting fed energy back into itself in a reinforcing loop until the oscillations grew violent enough to rip the structure apart. The wind wasn't even especially strong. The failure came from how the parts interacted as a system, a resonance that emerged from the deck's shape and flexibility combined with the wind. You could have tested every piece of that bridge to destruction in a lab and found nothing wrong, because the flaw wasn't in any piece. It was in the whole.

(The Tacoma Narrows Bridge illustrates a failure of systems engineering because the design focused on the strength of individual components but failed to adequately account for the bridge's emergent aerodynamic behavior as a complete system interacting with its environment.)

Emergence has a financial cousin that shows up in markets full of automated trading. On May 6, 2010, the U.S. stock market suffered what's now called the **Flash Crash**. In a matter of minutes, the Dow Jones Industrial Average plunged about a thousand points, nearly nine percent, and then largely recovered within the same afternoon. No single piece of news caused it and no single trader broke. What happened was that hundreds of automated trading systems, each running its own sensible rules, interacted in a way none of their designers had foreseen. A large sell order triggered algorithmic responses, those responses triggered others, liquidity evaporated as the machines pulled back from each other, and prices went into a brief free fall. Each algorithm did exactly what it was built to do. The crash was an emergent behavior of the whole market of interacting machines, a behavior that existed at the system level and nowhere in any individual program.

Why can't we just predict this behavior in advance and design it away? Because emergence comes precisely from the interactions, and the number of possible interactions in a complex system is vastly larger than the number of parts. A system with a few hundred components doesn't have a few hundred ways to behave; it has an astronomically large number of possible interaction patterns, and most of them are impossible to enumerate, let alone test. You can analyze every component thoroughly and still be blindsided, because the behavior you're worried about isn't a feature of any component. This is the deepest reason systems engineering can't be reduced to careful component engineering. The thing that bites you lives in a place that component analysis can't see.

Not all emergence is a threat, which is worth holding onto so the idea doesn't read as purely ominous. The useful function of almost any system is itself emergent. A car gets you across town, but no single part of a car can do that; transportation emerges from thousands of parts working together. A phone connects you to anyone on Earth, a capability that exists in no chip or wire alone. Designers often want emergence and work to produce it. The discipline of systems engineering, in one honest framing, is largely the practice of encouraging the emergent behaviors you want, like a working aircraft, while hunting down and preventing the emergent behaviors you don't, like a bridge that shakes itself to death. Both kinds are real, both come from the same source, and you can't have the good kind without accepting that the bad kind is always lurking in the same place.

Boundaries, Interfaces, and Environment

The last foundational idea answers a question that sounds almost too basic to ask: where does the system stop? Every system has to have a boundary, a line you draw that says this is part of the system and that is outside it, and where you draw that line shapes everything you do afterward. The **system boundary** is that line, and choosing it well is one of the first real decisions a systems engineer makes.

Consider designing a home thermostat. Is the system just the thermostat box on the wall? Or does it include the furnace it controls? The house's insulation, which determines how fast heat escapes? The electrical grid that powers it? The people who set it and open windows and complain when they're cold?

Each boundary you choose defines a different problem. Draw it tight around the box and you'll design a fine thermostat that performs poorly in a drafty house, because the house was outside your boundary so you never accounted for it. Draw it too wide and you'll drown trying to design the entire home energy situation when you were only asked for a thermostat. There's no universally correct boundary. There's only the boundary that fits the problem you're actually solving, and choosing it badly guarantees trouble that's hard to trace back to the choice.

Everything outside the boundary is the **environment**, the surroundings the system has to function within but doesn't control. The environment matters enormously, because a system that works in one environment can fail completely in another, and the environment is full of conditions the designers don't get to set. A satellite's environment includes vacuum, radiation, and temperature swings of hundreds of degrees between sunlight and shadow, all of which would destroy electronics built for a comfortable office. A medical implant's environment is the inside of a human body, warm, wet, salty, and chemically hostile to most materials. Software's environment includes the other programs on the machine, the network it depends on, and the users who do unexpected things. Defining the environment honestly, including the harsh and unlikely conditions, is part of defining the system, because the gap between the environment you designed for and the environment you actually get is where systems fail.

The piece that ties this section to the whole chapter, and arguably to the whole discipline, is the **interface**, the connection across a boundary where two things meet and exchange something, whether that's data, power, force, fluid, or signals. Interfaces are where systems engineering earns its keep, because interfaces are where independently built things have to agree, and agreement across a boundary is exactly what's hardest to guarantee when different teams own the two sides. Inside a single component, one engineer controls everything and can keep it consistent. At an interface, two separate parties have to share an identical understanding of what crosses the line, and if their understandings differ even slightly, the parts won't work together no matter how good each one is.

The cost of getting an interface wrong is best shown by the **Airbus A380**, the double-deck airliner that's among the most complex products ever manufactured.

During its development in the mid-2000s, Airbus hit a problem that delayed the entire program by about two years and cost billions of euros. The cause was an interface mismatch between design teams in different countries. The team in Germany was using one version of the CATIA design software, while the team in France was using a newer version, and the two versions handled the data differently. The aircraft's wiring is enormous, hundreds of miles of cable bundled into harnesses that thread through the airframe. When the German-designed harnesses were brought together with the French-designed structure, they didn't fit. The wires were the wrong length to reach where they needed to go. Both teams had done excellent engineering. Both had built precisely to their own understanding. But their understandings of the shared interface didn't match, because the tools that defined that interface weren't speaking the same language, and the result was a multibillion-euro failure in one of the most carefully run companies on Earth.

That story carries the central message of this whole chapter, which is why it sits at the end. The A380 wiring didn't fail because anyone built a bad part. It failed because two well-built parts met at a boundary where their understandings diverged, and the divergence was invisible until the physical pieces came together and refused to fit. This is the recurring pattern behind so much of what systems engineering deals with. The components are fine. The trouble is in the seams, the connections, the places where one party's assumptions meet another's and the two don't quite line up.

Because interfaces are where systems break, systems engineers manage them with a formality that can look excessive until you've watched an interface failure. They write **interface control documents** that nail down exactly what crosses each boundary: the format of the data, the voltage on the wire, the size and threading of the bolt, the units of the measurement, the timing of the signal. These documents exist so that two teams who may never meet, who may work in different companies, countries, and languages, can each build their side with confidence that the sides will fit. When the document is precise and both teams honor it, the parts come together cleanly. When the document is vague, or one team quietly deviates, or two versions of a tool interpret it differently, you get an A380 wiring problem, and you get it late, when it's most expensive to fix.

Standard interfaces are what let separately built things connect at all, and they're so common we stop noticing them. A USB port works because everyone agreed on exactly what the connector looks like, what the pins carry, and how the data flows, so a keyboard from one company plugs into a laptop from another and simply works. The electrical outlet in your wall is a defined interface that lets any compliant appliance draw power. These standards are nothing but interface agreements written down and adopted widely, and they're the reason a modern technological society can function with millions of parts built by millions of strangers who never coordinate directly. Every one of them rests on a boundary, an interface across it, and a shared, exact understanding of what crosses. Get that understanding right and complex systems hold together. Get it wrong and they come apart at the seams, every time, no matter how good the parts.

Discussion question: If a system can produce harmful behavior that none of its parts possess and that no one can fully predict by examining the parts, then who is responsible when that behavior causes damage, and on what basis can we hold them accountable for something no individual designed or could have reasonably foreseen?

Chapter 2: Requirements Engineering

There's an old rule of thumb in engineering that the cheapest way to build the wrong thing is to build it perfectly. You can assemble a brilliant team, write clean code, machine every part to tolerance, and deliver on schedule, and none of it matters if the thing you built doesn't solve the problem the customer actually had. Requirements engineering is the work of figuring out what to build before anyone builds it, and writing it down clearly enough that the people doing the building, the testing, and the buying all mean the same thing when they read it.

This is unglamorous work, and it's tempting to rush. Requirements live in documents and conversations, not in anything you can hold, and there's always pressure to skip ahead to something visible. But the front of a project is where the leverage is. A misunderstanding caught in a requirements review costs a meeting. The same misunderstanding caught after the system is built can cost a recall, a redesign, or a program. This chapter covers how you draw requirements out of the people who have them, how you sort the different kinds, how you keep track of where each one came from and where it goes, how you write them so they can actually be checked, and how you cope when they change underneath you, which they always do.

Eliciting Requirements from Stakeholders

Elicitation is the work of drawing requirements out of stakeholders, and the word is chosen carefully. You don't collect requirements the way you collect leaves, as if they're lying around fully formed waiting to be picked up. You draw them out, often against resistance, from people who don't fully know what they need, can't always say what they do know, and sometimes don't agree among themselves. The hard part of elicitation isn't writing down what people tell you. It's discovering what they couldn't tell you.

Three problems make this difficult, and they show up on nearly every project. The first is that stakeholders often don't know what they want until they see something concrete. Ask someone to describe their ideal software in the abstract and you'll get vague wishes. Show them a rough version and suddenly they have sharp, specific reactions: that button's in the wrong place, this step should come first, I didn't realize I'd need to do that twice a day. The knowledge was there, but it only surfaced when there was something to react to. The second problem is **tacit knowledge**, the things stakeholders know so well they've forgotten they know them. An experienced dispatcher or nurse or machinist has hundreds of small rules and exceptions running in their head, learned over years, that they never think to mention because to them it's just how the work is done. That hidden knowledge is exactly what the system has to honor, and it's exactly what stakeholders forget to say. The third problem is conflict. Different stakeholders want different things, and a requirement that satisfies one may violate another, so elicitation isn't only discovery, it's the start of negotiation.

Good elicitation uses several techniques together, because no single one catches everything. Interviews are the obvious starting point, one-on-one conversations where you ask open questions and listen hard for what's said and what's skipped.

Workshops bring stakeholders into a room together, which is useful precisely because it surfaces the disagreements early, while they're cheap to resolve, rather than late, when they've hardened into built systems that contradict each other. Observation, sometimes called watching the work, means going to where the work actually happens and watching people do it, because what people do and what they say they do are often different, and the gap is full of requirements. Document analysis means reading the forms, reports, regulations, and existing systems that already shape the work. And prototyping means building something rough and crude precisely so stakeholders have something to react to, which turns their vague wishes into specific corrections.

The technique that beginners underuse and experts rely on is observation, and there's a hard lesson behind why.

In London, the ambulance service tried to replace its manual dispatch system with a computer-aided one, and it went live on October 26, 1992. Within hours it was in trouble. The system was supposed to track ambulances and route them to emergencies faster than human dispatchers could, but it couldn't keep up with how the real work actually flowed. Ambulances got lost in the system, calls stacked up, the same incident sometimes generated multiple dispatches, and response times collapsed. The service fell back to manual methods within days, and the failure became one of the most studied disasters in software history. The official inquiry was blunt about the cause. The people who specified the system hadn't understood how dispatching actually worked on the ground. They'd built a system around an idealized model of the process, not the messy reality that experienced dispatchers navigated, with all its workarounds, judgment calls, and exceptions. The dispatchers' tacit knowledge, the stuff that made the manual system function, was never elicited, because nobody spent enough time watching them do the job. The requirements described a dispatch process that didn't match the one in the building. Each requirement may have been written correctly. They were just requirements for the wrong system.

That's the deepest danger in elicitation, and it's worth naming clearly: you can do everything right at every later stage and still fail completely if the requirements describe a system that solves the wrong problem. No amount of good design, clean code, or thorough testing recovers from requirements that captured the wrong need. The error gets baked in at the start and faithfully reproduced all the way to delivery. This is why elicitation, the least technical part of engineering, is often the part that decides whether a project lives or dies.

There's a skill in elicitation that separates good analysts from order-takers, and it's the habit of digging past the stated request to the underlying need. When a stakeholder asks for a specific feature, that request is usually a solution they've imagined to a problem they haven't described. A manager who demands a daily report at 8 a.m. might really need to spot a particular problem before a morning meeting, and there may be a better way to surface that problem than a report nobody reads. If you build exactly what people ask for without understanding why they're asking, you'll faithfully implement their guesses about how to solve their problems, guesses that are often worse than what an engineer who understood the real need would design. The question "why do you need that?" asked gently and repeatedly, is one of the most productive tools in the discipline.

Elicitation never produces a clean, complete, consistent set of requirements on the first pass, and expecting it to is itself a mistake. What you get is partial, contradictory, and full of gaps you won't notice until later. Good practice treats elicitation as iterative: you draw out what you can, write it down, show it back to stakeholders, watch them react and correct, and go around again. Each loop sharpens the picture. The stakeholders who said one thing in the first interview will say something different once they've seen the requirements written down, not because they're being difficult, but because seeing it concrete revealed what they actually meant. The job isn't to get it right the first time. It's to converge on the truth through enough honest loops that the gaps close before they become expensive.

Functional Versus Non-Functional Requirements

Once you're drawing requirements out, you'll notice they come in two flavors, and the distinction matters because teams handle one well and the other badly.

Functional requirements describe what the system does, the specific behaviors and functions it performs. **Non-functional requirements** describe how well it does those things, the qualities the system has to have while performing its functions. The difference sounds academic until you see how often the second kind gets neglected, and what neglecting it costs.

An automated teller machine makes the distinction concrete. The functional requirements are the things the machine does: it verifies the customer's PIN, it dispenses the requested amount of cash, it accepts deposits, it prints a receipt, it transfers funds between accounts. Each describes an action the system performs, a function it carries out. You can point at each one and watch the machine do it.

The non-functional requirements are different in kind. The ATM should complete a cash withdrawal within ten seconds of the PIN being verified. That's not a function; it's a quality, in this case speed. The machine should be available 99.9 percent of the time, a quality of reliability. It should encrypt every message it sends to the bank, a quality of security. Its screen should be readable in direct sunlight, a quality tied to its environment. None of these describes something the ATM does in the sense of a function. They describe constraints on how it does everything, and the machine can perform every one of its functions perfectly while failing every one of these qualities, in which case it's a fast, capable, useless machine that customers can't trust.

Non-functional requirements come in recognizable categories, and knowing the categories helps you remember to ask about all of them rather than fixating on whichever ones come to mind first. Performance covers speed and responsiveness, how fast the system reacts and how much it can handle. Reliability and availability cover how often the system fails and how much of the time it's working. Security covers protection against attack and unauthorized access. Usability covers how easily people can actually use the thing. Maintainability covers how hard it is to fix and update. Scalability covers whether the system keeps working as demand grows. Compliance covers whether it meets the laws and standards it's subject to. These qualities cut across the whole system rather than attaching to any single function, which is part of why they're easy to overlook: there's no one feature to point at, so they slip through the cracks between features.

Here's why the distinction earns its own section. Functional requirements get attention naturally, because they're what the customer asks for and what the demo shows off. Non-functional requirements get neglected, because they're invisible until they're violated. Nobody notices that a system is secure or reliable; they only notice when it's breached or down. So teams pour effort into making the system do impressive things and treat the qualities as an afterthought, and then the system ships fast and full-featured and falls over under real load, or gets hacked, or turns out to be unusable by the people it was built for. The non-functional requirements are frequently the ones that determine whether a system survives contact with reality, and they're frequently the ones written last, vaguely, or not at all.

The cost of neglecting a non-functional requirement, when that requirement is safety, can be lethal. The **Therac-25** was a radiation therapy machine used to treat cancer patients in the mid-1980s. Between 1985 and 1987, it gave at least six patients massive radiation overdoses, hundreds of times the intended dose, and several of them died. The machine's functional requirements were met: it delivered radiation to the target as instructed. What failed were the qualities, in particular safety. Earlier versions of the machine had hardware interlocks, physical safeguards that made certain dangerous states impossible. The Therac-25 removed many of those and relied on software to prevent unsafe conditions instead, but the software safety requirements were never specified or verified with the seriousness that life-critical behavior demands. Under a specific, unusual sequence of operator keystrokes, the software entered a state that allowed a lethal dose, and there was no longer a hardware safeguard to catch it. The people who designed it weren't careless about whether the machine could deliver radiation. They were careless about the quality of safety, treating it as something the software would handle rather than something that had to be specified, demanded, and proven. The functional requirements were fine. The neglected non-functional requirement killed people.

Non-functional requirements are also harder to pin down, which is part of why they get shortchanged, and it connects to the next two sections. It's easy to write "the system shall print a receipt," and easy to check whether it does. It's harder to write a good requirement for "the system shall be reliable," because reliable doesn't mean anything you can test until you turn it into a number: reliable how, measured how, to what standard? A functional requirement is often a yes-or-no question. A non-functional requirement usually demands a measurement, a threshold, and a method for checking it, which is more work to write and more work to verify. Teams under time pressure write the easy ones well and the hard ones badly, and the hard ones are often the ones that matter most.

Requirements Traceability

As a system grows, the number of requirements grows with it, and a serious system can have thousands, scattered across documents and owned by different teams. **Traceability** is the practice of maintaining clear links that connect each requirement backward to its source and forward to whatever satisfies and verifies it. The idea is simple. The discipline of actually doing it is what separates projects that can answer "why does this exist?" from projects that can't.

Trace links run in two directions, and both matter. **Backward traceability** connects a requirement to where it came from: the stakeholder need, the regulation, the higher-level requirement that gave rise to it. If you can trace backward, you can answer the question "why is this requirement here?" for every requirement in the system, and you can spot requirements that trace back to nothing, which usually means somebody added a feature that no real need justified. **Forward traceability** connects a requirement to what implements it and what tests it: the design element that satisfies it, the component that contains it, the test that proves it's met. If you can trace forward, you can answer "where is this requirement satisfied, and how do we know it works?" for everything the system is supposed to do, and you can spot requirements that nothing implements or nothing tests, which are requirements that will quietly go unmet.

The tool that holds this together is the **requirements traceability matrix**, a structured map linking each requirement to its source, its design, and its verification. It sounds bureaucratic, and maintaining it is real work, but it answers questions that are nearly impossible to answer otherwise. For example:

REQ ID	REQUIREMENT	SOURCE	DESIGN ELEMENT	TEST CASE	STATUS	COVERED
REQ-01	System shall authenticate users within 2 seconds	SRS §3.1	AuthService.login()	TC-AUTH-01	pass	yes
REQ-02	User data shall be encrypted at rest (AES-256)	SEC-POL-4	EncryptionModule	TC-SEC-03	pass	yes
REQ-03	System shall support 10,000 concurrent users	SRS §4.2	LoadBalancer config	TC-PERF-02	pending	yes
REQ-04	Audit log shall capture all admin actions	REG-HIPAA	AuditLogger	TC-AUD-01	pass	yes
REQ-05	Password reset flow shall expire tokens in 15 min	SRS §3.4	TokenService.expire()	—	no test	no
REQ-06	Export reports in PDF and CSV formats	CR-2024-11	ReportExporter	TC-EXP-01	pending	yes

pass verified pending test written, not run no test gap — not covered

(REQ-05 is the telling row – a requirement with a source, a design element, but no test case. That's exactly the gap the matrix surfaces: the requirement exists, someone designed to it, but there's no verification. Without the matrix, that hole stays invisible until a bug report or an audit finds it.

The two expensive questions it answers: "does every requirement have a test?" (coverage, shown in the last column) and "why does this test exist?" (traceability backward to the source, useful when someone wants to cut scope).)

Two of those questions come up constantly on real projects, and getting them wrong is expensive.

The first is impact analysis. When something changes, and something always changes, you need to know what else is affected. A customer revises a requirement, a regulation shifts, a design constraint turns out to be infeasible. Without traceability, you're guessing about the ripple effects, and you'll miss some, and the ones you miss become bugs and failures that surface much later, far from their cause. With a traceability matrix, you can follow the links and see exactly which design elements, components, and tests are touched by a change, so you can update all of them on purpose instead of discovering the misses in the field. On a large system, this is often the difference between a controlled change and a cascade of broken assumptions.

The second is coverage. You need to be able to prove, not hope, that every requirement is actually implemented and tested, and that nothing's been built that no requirement asked for. Traceability gives you both directions: every requirement should trace forward to a test, and every test should trace back to a requirement. A requirement with no test linked to it is a promise nobody's checking. A component with no requirement behind it is effort spent on something nobody needed, which at best is waste and at worst is unrequested behavior hiding in the system.

In regulated industries, traceability isn't optional, it's law, because the consequences of unverified requirements are severe enough that society demands proof. Aircraft software in the United States and Europe is governed by a standard called DO-178C, which requires complete traceability from requirements through design and code to test, so that for any line of behavior in a flight system there's a documented chain showing why it exists and how it was verified. Medical devices fall under FDA requirements that demand similar traceability. Automotive safety systems follow ISO 26262, which traces safety requirements through the design. These rules exist because regulators learned, often from disasters, that systems built without traceable requirements tend to have requirements that go unverified, and unverified requirements in an aircraft or a pacemaker or a car's braking system are how people die.

The cost of a broken trace chain showed up tragically in the **Boeing 737 MAX**. The aircraft included a system called MCAS, the Maneuvering Characteristics Augmentation System, designed to push the nose down automatically under certain conditions to make the new, larger-engined plane handle like its predecessors. Two of these aircraft crashed within five months, Lion Air Flight 610 in October 2018 and Ethiopian Airlines Flight 302 in March 2019, killing 346 people between them. The investigations that followed found, among other failures, a breakdown in how the system's requirements traced through to its safety analysis. As MCAS was developed, its authority grew: it ended up able to command larger and more repeated nose-down movements than early versions, and it relied on a single angle-of-attack sensor with no backup. But the safety assessment that should have evaluated the consequences of that authority and that single point of failure didn't fully reflect what the system had become. The chain that should have linked the system's actual behavior to a rigorous analysis of what could go wrong had gaps. When the sensor fed bad data, MCAS pushed the nose down repeatedly, and the pilots couldn't recover. The requirements existed. The aircraft was built. But the trace from what the system could do to whether that was safe had broken, and the consequence was two aircraft full of people.

Traceability is tedious to maintain, and on a small project it can feel like overhead that slows you down for no visible gain. On a large or safety-critical project, it's the connective tissue that lets you answer the questions that matter: why does this exist, where is it satisfied, how do we know it works, and what breaks if we change it. Projects that skip it don't notice the cost until they need answers and don't have them, usually at the worst possible moment, when a change has to be made fast or a failure has to be traced under pressure.

Writing Verifiable Requirements

A requirement that can't be checked isn't a requirement; it's a wish dressed up as one. **Verifiable** means you can determine, by some objective method, whether the requirement is met. This is the single most useful test to apply to any requirement, and applying it ruthlessly catches more bad requirements than any other technique. If you can't describe how you'd prove a requirement is satisfied, the requirement is broken, no matter how reasonable it sounds.

The problem is that unverifiable requirements feel completely fine when you write them. "The system shall be user-friendly." "The interface shall be intuitive." "The system shall be fast." "The application shall be highly reliable." "Errors shall be handled appropriately." Every one of these sounds like a real requirement, and every one is useless, because none can be objectively checked. User-friendly to whom, measured how? Fast meaning what, under what load? Highly reliable to what standard? Appropriately according to whose judgment? These requirements will be interpreted differently by everyone who reads them, the engineer will build to one interpretation, the customer will expect another, and the disagreement will surface at acceptance, when it's expensive and adversarial. The requirement that means something different to each reader is worse than no requirement at all, because it creates the illusion of agreement where none exists.

Turning a wish into a verifiable requirement means replacing the vague quality with something measurable. Watch the transformation:

"The system shall be fast" becomes "The system shall return search results within 2 seconds for 95 percent of queries when serving 10,000 concurrent users." Now there's a number, a condition, and a threshold, and you can run a test that gives a clear pass or fail.

"The website shall be user-friendly" becomes "A first-time user shall complete the checkout process in under 90 seconds without assistance, verified through usability testing with at least 20 representative users." User-friendly was an opinion. This is a measurement.

"The system shall be highly available" becomes "The system shall maintain 99.95 percent uptime measured monthly, excluding maintenance windows announced 48 hours in advance." Highly available meant nothing. This means about 22 minutes of allowed downtime per month, and either you hit it or you don't.

The pattern is consistent: find the vague quality word, ask "how would I measure that?", and rewrite the requirement around the measurement. If you can't find a way to measure it, that's a sign the requirement isn't ready, and you need to go back to the stakeholder and pin down what they actually mean before you write anything down.

There's a writing convention that's spread across the field for good reason: the word **shall** marks a binding requirement. When a requirements document says "the system shall," it means this is mandatory, not a suggestion. The convention distinguishes shall from related words: should marks a goal or recommendation that's desired but not mandatory, will marks a statement of fact or intent rather than a requirement on the system, and may marks something optional or permitted. Keeping these straight matters because mixing them creates confusion about what's actually required. A document full of "the system should ideally try to" gives engineers no clear obligation and customers no clear promise. The discipline of writing real requirements as "shall" statements, and reserving the softer words for things that genuinely are softer, keeps the line between obligation and aspiration visible.

A well-written requirement has a handful of properties, and checking each requirement against them catches most defects before they cause trouble.

- **Necessary.** It traces to a real need. If you can't say why it's there, it shouldn't be.
- **Unambiguous.** It has one interpretation, not several. Different readers reach the same understanding.
- **Verifiable.** You can prove objectively whether it's met.
- **Feasible.** It can actually be done within the project's constraints.
- **Singular.** It states one requirement, not several bundled together, so it can be tracked and verified on its own.
- **Consistent.** It doesn't contradict other requirements.

That singular property catches a common mistake. People write requirements like "the system shall encrypt all data and respond within two seconds and log every transaction." That's three requirements wearing one sentence, and it causes problems: you can't trace or verify them separately, and if one part fails, the status of the whole statement is unclear. Each requirement should make one testable claim, so that each can be tracked, satisfied, and verified on its own. Splitting bundled requirements apart is dull work that prevents a surprising amount of confusion downstream.

The danger of imprecise requirements isn't theoretical, and one of the most famous examples in the history of spaceflight came down to a single mark on paper. In 1962, NASA launched **Mariner 1**, a probe meant to fly past Venus. Shortly after liftoff the rocket began to veer off course, and range safety destroyed it less than five minutes into flight, a loss of around 18 million dollars in the money of the time. The investigation traced the failure to the guidance software, and the software error traced to the specification it was built from. In the handwritten guidance equations, a symbol had been written incorrectly: a bar that was supposed to sit over a variable, indicating a smoothed or averaged value, was missing. The programmer faithfully coded what the specification said, and what it said, with the bar missing, told the rocket to respond to normal variations as if they were errors to correct, so it chased phantom problems until it pointed the wrong way. The popular telling calls it the most expensive hyphen in history, which oversimplifies the symbol involved, but the lesson holds exactly. A specification that's even slightly wrong gets faithfully turned into a system that's wrong, because the people downstream build what the requirement says, not what its author meant. Precision in writing requirements isn't fussiness. It's the difference between a probe that reaches Venus and one that's blown up over the Atlantic.

Writing verifiable requirements is a craft, and like any craft it improves with the discipline of always asking the same hard question: how would I prove this is met? Applied to every requirement, that question forces out the vagueness, the bundled claims, the wishful qualities that can't be checked. It's slower than dashing off "the system shall be reliable and easy to use," and it produces requirements that are longer, drier, and full of numbers. It also produces requirements that mean the same thing to the engineer who builds, the tester who checks, and the customer who pays, which is the entire point of writing them down.

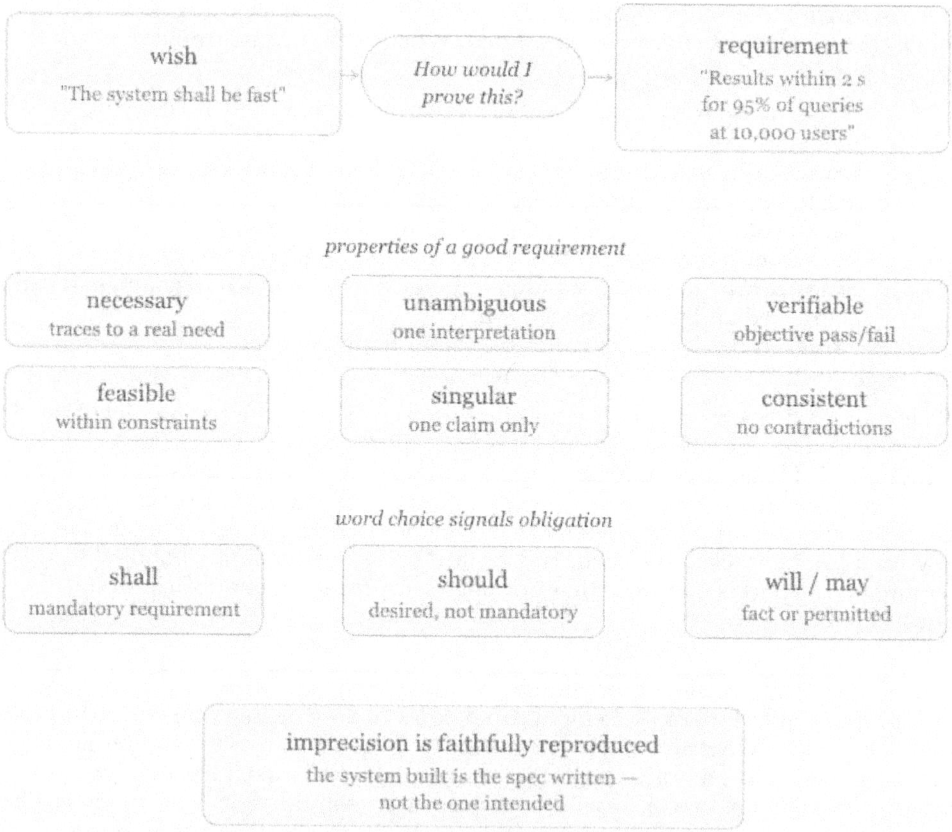

Managing Requirements Volatility

Requirements change. Not occasionally, not on badly run projects, but on essentially every project of any size, all the time. **Requirements volatility** is the rate and extent to which requirements change over the life of a project, and pretending it won't happen is one of the surest ways to be wrecked by it. The goal isn't to stop change, which is impossible, but to manage it so that necessary changes get made in a controlled way and unnecessary ones get stopped.

Requirements change for honest reasons. Stakeholders learn what they actually need once they see early versions, which means the act of building the system reveals requirements that elicitation missed. The environment shifts: competitors release new products, the business strategy changes, the market moves. Regulations change, forcing the system to comply with new rules. Technology changes, opening options that didn't exist when the requirements were written. Stakeholders themselves change as people leave and new ones arrive with different priorities. None of this is anyone's fault. It's the nature of building something over months or years in a world that doesn't hold still. A project that treats every change as a failure of planning will exhaust itself fighting reality.

But there's a darker source of change that has to be distinguished from the honest kind: **scope creep**, the slow, uncontrolled expansion of what the system is supposed to do, one small addition at a time. Each individual addition seems reasonable. Could we also add this? While we're at it, what about that? It would be nice if it also did this other thing. No single request is unreasonable, and saying no to any one of them feels petty. But they accumulate, the system grows beyond what the schedule and budget can support, and the project that was on track for a focused, achievable goal collapses under the weight of a hundred small yeses. Scope creep is dangerous precisely because it's invisible in the moment: there's no single decision to expand the project, just a steady drift that nobody chose and nobody can point to, until the deadline arrives and nothing's finished.

The mechanism that separates controlled change from chaos is a **change control process**, often run through a **change control board**. The idea is straightforward. Once requirements are agreed and written down, they're frozen into a **baseline**, a documented, approved version that everyone works from. After that, changes don't happen because someone asks in a hallway. A proposed change goes through a defined process: it's written down, its impact is analyzed, this is where traceability earns its keep, and a board with the authority to decide weighs the benefit against the cost in schedule, budget, and risk. If it's worth it, the change is approved and the baseline is formally updated. If it isn't, it's declined. The point isn't to make change hard for its own sake. It's to make sure every change is a decision somebody made on purpose, with knowledge of what it costs, rather than a drift nobody controlled.

What uncontrolled requirements volatility does to a project is illustrated by one of the most expensive software failures the U.S. government ever owned. After September 11, 2001, the FBI tried to build a modern case management system called the Virtual Case File, part of a larger modernization program. The bureau spent roughly 170 million dollars, and in 2005 the system was scrapped without ever being deployed. There were several causes, but a central one was requirements that wouldn't hold still. The bureau kept changing what it wanted, the requirements were never frozen into a stable baseline, and the contractor was building against a target that moved faster than the work could keep up with. By the time one set of requirements was being implemented, the requirements had changed again, and the system never converged on anything finishable. The money bought a great deal of effort and no working system, because the foundation, a stable agreed statement of what to build, never solidified. Volatility that isn't managed doesn't just slow a project. It can prevent the project from ever finishing at all.

Late changes are especially destructive, and a famous case shows how a single ambitious requirement, pushed too late and underestimated, can sink a project's schedule. Denver built a new international airport in the early 1990s, and the plan included an automated baggage handling system meant to move luggage through the entire airport without human carriers, a system far more ambitious than anything built before. The complexity was badly underestimated, and changes and additions kept coming as construction proceeded. The baggage system couldn't be made to work reliably: bags were lost, jammed, and mangled in testing.

The problems delayed the airport's opening by about sixteen months, the airport finally opened in February 1995 instead of October 1993, and the overruns ran into hundreds of millions of dollars. Eventually most of the automated system was abandoned in favor of conventional methods. Part of what made it so costly was timing. The ambition and the changes collided with a system that was already being built, so every problem had to be solved on physical hardware already installed, at the most expensive point in the project to be discovering that the requirements were harder than anyone admitted.

There's a spectrum of approaches to handling volatility, and choosing the right one depends on how much change you expect. Plan-driven methods, often called waterfall, try to get requirements as complete and stable as possible up front, then build against them with tight change control. This works when requirements are well understood and unlikely to shift much, as in some hardware and infrastructure projects where late change is genuinely ruinous. Agile methods take the opposite stance: they assume requirements will change, so they build in short cycles, deliver working pieces frequently, and expect to revise direction based on what each cycle reveals. This works when requirements are uncertain and feedback is the fastest way to discover the real need, as in much software development. Neither is universally right. A project that uses rigid up-front planning where requirements are genuinely uncertain will lock in the wrong requirements and struggle to change them. A project that uses loose iterative methods where requirements are stable and the cost of late change is enormous, like a bridge or a satellite, invites disaster.

Matching the approach to the actual volatility of the requirements, and to the cost of changing course late, is part of the judgment that requirements engineering demands.

Discussion question: If stakeholders genuinely can't know what they want until they see and use a working system, then how can any requirement written before that system exists be trusted, and where should a team draw the line between specifying enough up front to build coherently and accepting that the real requirements will only be discovered through building?

Chapter 3: System Architecture and Design

Around 2002, Amazon's leadership made a decision that looked like an internal plumbing change and turned out to reshape the company. The rule, issued in a now-famous memo, was that every team had to expose its data and functionality through service interfaces, and teams could communicate with each other only through those interfaces. No shared databases, no reaching into another team's code, no back doors. If your team needed something another team had, you called their service the same way an outside customer would. People grumbled, because it was harder in the short term. But that single architectural choice forced Amazon to build clean, well-defined boundaries between every part of its system, and once those boundaries existed, the company realized it could sell access to them. The infrastructure that became Amazon Web Services, the business that now earns more profit than Amazon's entire retail operation, grew out of an architecture decision about how the pieces should talk to each other.

That's what architecture is, and why it gets its own chapter before any detailed design. **Architecture** is the set of high-level decisions about how a system is structured: what the major pieces are, how they're arranged, and how they connect. These decisions get made early, when you know the least, and they're the hardest of all to change later, because everything else gets built on top of them. A detail can be fixed in an afternoon. An architecture is woven through the entire system, and changing it means unweaving and reweaving the whole thing.

This chapter covers how you choose an architecture from the patterns available, how you break a system into pieces that can be built separately, how you write down the agreements that hold those pieces together, how you design for reliability and for manufacturing from the start, and how the discipline is shifting from documents to connected digital models.

Architectural Patterns and Trade-offs

Architects don't invent a fresh structure for every system, any more than engineers reinvent the bolt. Over decades, certain arrangements have proven useful enough to be named, studied, and reused, and these are **architectural patterns**: recurring structural solutions to common design problems. Knowing the patterns means you start from accumulated experience instead of a blank page, and you can reason about what each one will give you and cost you before you commit.

A few patterns show up across nearly every domain. A **layered architecture** stacks the system in horizontal levels, where each layer uses the one below it and serves the one above. Operating systems work this way, with hardware at the bottom, then a kernel, then system services, then applications on top. The benefit is that each layer can change internally without disturbing the others, as long as the boundary between layers stays stable. A **client-server architecture** splits the system into clients that request things and servers that provide them, which is how the web works: your browser is a client, and somewhere a server answers it. A **pipeline architecture** passes data through a sequence of processing stages, each transforming it and handing it on, which suits compilers, data processing, and audio and video workflows. A **monolithic architecture** builds the whole system as a single unified unit, while a **microservices architecture** breaks it into many small, independent services that each do one thing and communicate over a network, which is the pattern Amazon's mandate pushed the company toward.

The choice between monolithic and microservices is worth dwelling on, because it shows the nature of architectural decisions clearly: there's no right answer, only a set of trade-offs you accept. A monolith is simpler to build at first, easier to test as one piece, and faster to run because everything's in one place with no network between the parts. Its weakness shows up as it grows. A large monolith becomes hard to change, because everything's connected to everything, and a small modification can ripple through the whole system in ways nobody fully tracks. You also have to deploy the entire thing at once, so a tiny fix means redeploying everything, and one team's work can block another's.

Microservices flip every one of these. Each service is small, owned by one team, deployable on its own, and changeable without touching the rest. The cost is that you've traded the complexity inside a monolith for complexity between services: now you have a network in the middle that can fail, dozens of independent pieces to coordinate, and the genuinely hard problem of making many separate services behave correctly as a whole. Companies like Netflix and Amazon moved to microservices because their scale made monoliths unmanageable, and the trade was worth it for them. A small startup that copies the microservices pattern because the big companies use it often drowns in the coordination overhead before it ever reaches the scale that would justify the choice.

This is the central truth of architecture, and it's worth stating plainly: every architectural pattern optimizes some qualities at the expense of others, and there is no arrangement that's best at everything. An architecture tuned for performance often sacrifices flexibility. One tuned for flexibility often sacrifices simplicity and speed. One tuned to be cheap to build is often expensive to change. The architect's job isn't to find the perfect structure, which doesn't exist, but to understand which qualities matter most for this particular system and choose the pattern whose trade-offs serve those qualities. A system where reliability matters above all, like flight control, makes different architectural choices than a system where rapid feature development matters most, like a consumer app, and both are correct for their context.

The same trade-off logic governs a choice that runs deeper than software patterns: whether to centralize or distribute. A **centralized architecture** concentrates control and processing in one place, which is simpler to manage, easier to keep consistent, and cheaper to coordinate, but creates a single point whose failure takes down everything and a single bottleneck that limits how much the system can grow. A **distributed architecture** spreads control across many locations, which removes the single point of failure and can scale much further, but is far harder to keep consistent and coordinate, because now many parts have to agree without a central authority telling them what's true. Power grids wrestle with exactly this. The traditional grid was centralized, with a few large power plants feeding electricity outward to consumers, which was efficient and controllable. The grid is now shifting toward distributed generation, with rooftop solar, batteries, and wind feeding power in from thousands of points, which is more resilient against the failure of any one source but vastly more complicated to balance, because the operator no longer controls a handful of plants but has to coordinate a sprawling network of small, variable, independently owned sources. Neither grid is simply better. They make different trade-offs, and the industry is living through the difficulty of moving from one to the other.

There's a hidden force that shapes architecture whether or not anyone intends it, named after the engineer who described it in 1967. **Conway's Law** observes that organizations tend to design systems that mirror their own communication structure. If you have four teams that don't talk much, you'll tend to get a system with four parts and awkward connections between them, because the system's interfaces end up matching the gaps between the teams. This sounds like a curiosity but has real consequences. It means the way you organize your people shapes the architecture you'll get, often more than your deliberate design choices do. Smart organizations use this deliberately, sometimes called the inverse Conway maneuver: if you want a particular architecture, structure your teams to match it, so the communication

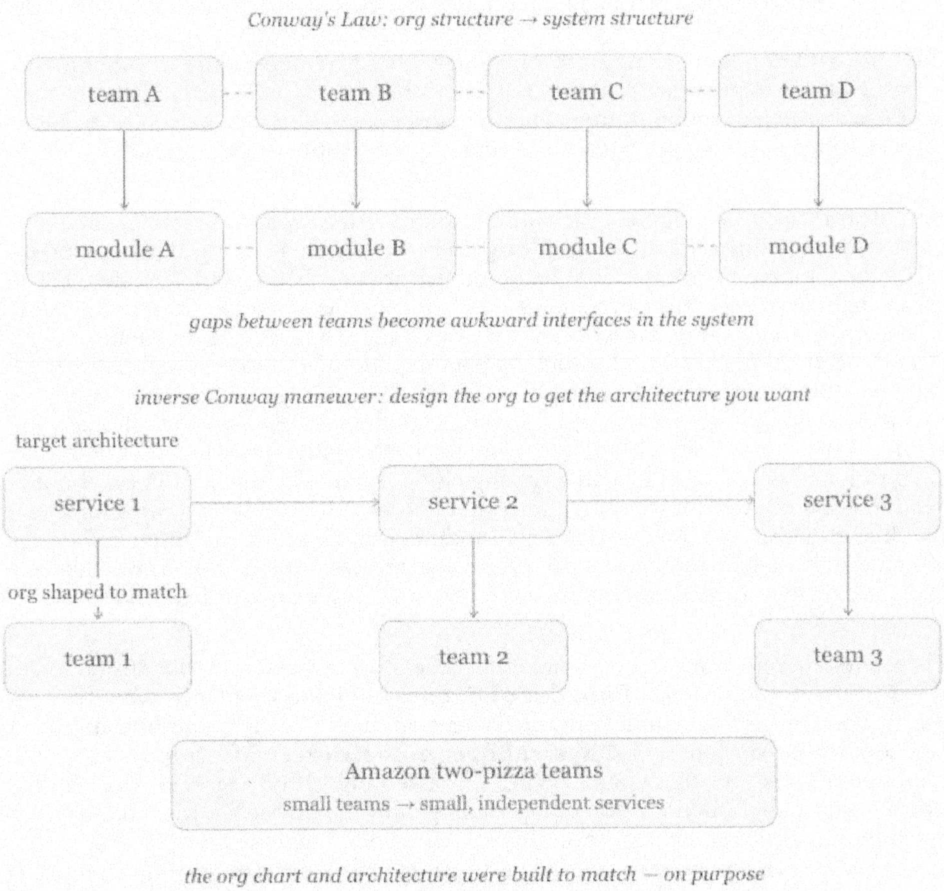

patterns push the design in the direction you want rather than fighting it. Amazon's two-pizza teams, small teams sized so two pizzas could feed them, weren't just a management fad; they were an organizational structure designed to produce the small, independent services the architecture called for. The architecture and the org chart were built to match, because Conway's Law guarantees they'll match anyway, so you might as well choose the match on purpose.

Modularity and Decomposition

Having chosen an architecture, you face the question of how to actually divide the system into pieces, and this is **decomposition**: breaking a large system into smaller parts that can be understood, designed, and built separately. The way you decompose a system shapes everything downstream, because the pieces you create become the units that teams own, that get built and tested independently, and that have to be reassembled into a working whole. Decompose well and the parts come together cleanly. Decompose badly and you spend the rest of the project fighting the boundaries you drew.

The quality you're aiming for is **modularity**, which means building a system from parts that are self-contained and interchangeable, each hiding its internal workings behind a clean interface. A module is modular when you can understand it, build it, test it, and replace it without needing to know the guts of every other module. This is what lets large teams work in parallel without constantly tripping over each other, and it's what lets you fix or upgrade one part without rebuilding the whole. The opposite of modularity is a system where everything's tangled together, where you can't touch one part without understanding and possibly breaking ten others, and where every change is risky because the consequences spread unpredictably.

Two ideas measure how good a decomposition is, and engineers reach for them constantly: coupling and cohesion. **Coupling** is how much the modules depend on each other, and you want it low. When modules are tightly coupled, a change in one forces changes in others, failures propagate between them, and you can't understand or test any module alone. **Cohesion** is how well the contents of a single module belong together, and you want it high. A cohesive module does one well-defined job, and everything in it serves that job. The combined goal, repeated across every engineering discipline, is low coupling and high cohesion: each module should be tightly focused internally and loosely connected externally. A module that does one thing well and barely depends on anything else is a module you can reason about, build in isolation, test thoroughly, and swap out later. Get this wrong, with high coupling and low cohesion, and you get modules that do scattered, unrelated things while depending heavily on each other, which is the worst of both worlds: you can't understand any module on its own, and you can't change any of them safely.

There are different ways to slice a system, and choosing the wrong dimension to cut along is a common mistake. **Functional decomposition** divides by what the system does, grouping by function: the navigation function, the communication function, the power function. **Physical decomposition** divides by physical components: this box, that circuit board, this assembly. Often a system needs both views, because the functions don't map cleanly onto the physical parts, and one function may be spread across several boxes while one box may serve several functions. The discipline is to be deliberate about which decomposition you're using

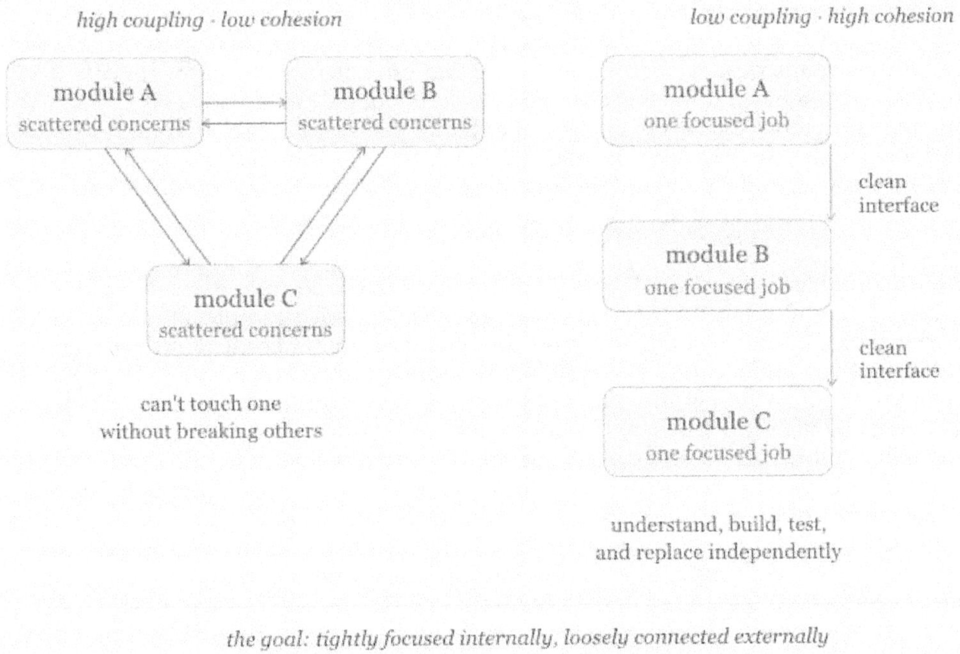

and why, rather than mixing them carelessly and ending up with parts that are neither functionally clean nor physically clean.

The payoff of good modular decomposition is best seen in a machine that defined an industry. In 1964, IBM announced the **System/360**, and it changed how computers were built. Before it, each computer model was its own design, and software written for one wouldn't run on another, so customers who outgrew a machine had to throw away everything and start over. IBM's idea was a family of compatible computers, from small and cheap to large and powerful, that all shared the same architecture, so the same software ran across the whole range. A customer could buy a small model and later move up to a larger one without rewriting anything.

This required ruthless modularity: the instruction set, the interfaces, and the architecture had to be defined cleanly enough that wildly different hardware could implement the same behavior, and peripherals like printers and storage could be designed once and attached to any model. The System/360 was an enormous bet, reportedly costing IBM more than it was worth at the time, and it succeeded so completely that its architecture's descendants are still running in mainframes today, sixty years later. The technical foundation of that success was modular decomposition done with unusual rigor: pieces with clean interfaces that could be mixed, matched, and upgraded independently.

The same principle drives modern cars, where it saves billions. Volkswagen built a modular platform called MQB, a shared set of components and interfaces that underpins an enormous range of vehicles, from the Golf to several Audi and Škoda models. The cars look different and serve different markets, but underneath they share a common architecture: the same basic structure, the same mounting points, the same approach to where the engine and major components sit. This modularity lets the company design a component once and use it across dozens of models, develop new vehicles faster because much of the foundation already exists, and manufacture more efficiently because the shared parts run at higher volume. The customer sees a distinct car. The engineering reality is a modular platform decomposed so that the shared pieces and the model-specific pieces have clean boundaries between them.

Modularity has limits, though, and pretending it's free leads to a particular kind of failure. Every interface between modules has a cost: it adds complexity, it can hurt performance because data has to cross the boundary, and it constrains the design because the modules on each side have to honor the agreed connection. Sometimes the most modular design isn't the best design, because the cost of all those clean boundaries outweighs the benefit. Google learned this with Project Ara, an attempt to build a fully modular smartphone where you could swap out the camera, battery, screen, and other parts like building blocks. The vision was appealing: upgrade one piece instead of buying a whole new phone, customize the device to your needs. But making everything modular meant every module needed connectors and interfaces, which added bulk, weight, cost, and points of failure, and the modular phone ended up worse than an integrated one on the things customers actually cared about, like size, performance, and price. Google cancelled the project in 2016. The lesson isn't that modularity is bad; it's that modularity is a trade-off like everything else in architecture, and there's such a thing as too much of it. The right amount of modularity depends on whether the flexibility it buys is worth the complexity it adds, and that balance is different for a mainframe, a car, and a phone.

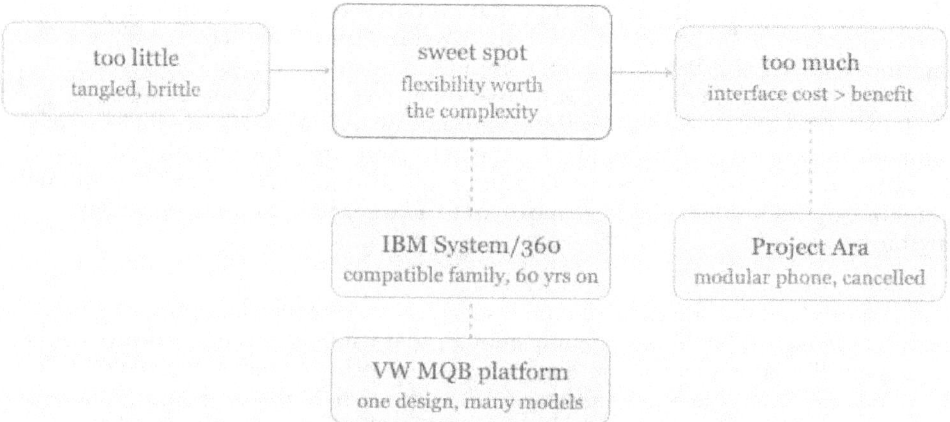

modularity is a trade-off: right amount depends on whether flexibility justifies complexity

Interface Control Documents

When two parts of a system are designed by different people, possibly in different organizations, the agreement about how they connect has to be written down with care, and that written agreement is the **interface control document**, or ICD. The earlier chapter established why interfaces are where systems break; this is about the discipline of actually defining them so they don't. An ICD is the contract between two sides of a boundary, specifying exactly what crosses it so that each side can build independently and trust that the pieces will meet.

Interfaces come in several types, and a thorough ICD addresses every type that applies, because a connection that's perfect on one dimension can still fail on another. A mechanical interface specifies physical fit: dimensions, mounting points, bolt patterns, clearances, the shape and size of mating surfaces. An electrical interface specifies power and signals: voltages, currents, connector pinouts, what each wire carries. A data interface specifies information exchange: message formats, protocols, data types, units, timing, what happens in what order. A thermal interface specifies how heat moves across the boundary, which matters enormously in tightly packed systems where one component's heat can cook its neighbor. There are others, fluid, optical, software, but the point holds across all of them: a single boundary often involves several kinds of interface at once, and the ICD has to nail down each one, because the parts can fit mechanically and still fail electrically, or match electrically and still mismatch on data format.

The defining feature of an ICD is that it's owned jointly and changed formally, never casually. The whole reason it exists is so that neither side can change the interface without the other knowing, because a one-sided change to a shared boundary is exactly how parts that were supposed to fit stop fitting. So an ICD is baselined like a requirement, frozen into an agreed version, and any change to it goes through a controlled process where both sides agree before anything moves. When one team quietly decides to use a slightly different connector or change a message format "just a little," and doesn't update the ICD or tell the other side, the result is two parts built to two different understandings of the same boundary, discovered at integration when it's expensive and adversarial to fix.

A vivid demonstration of an ICD done right under genuinely hard conditions came in 1975, when an American Apollo spacecraft and a Soviet Soyuz docked in orbit during the Apollo-Soyuz Test Project. The two countries had built their spacecraft independently, with no shared standards, in the middle of the Cold War, and they had to physically join in space and let crew pass between them. The interfaces involved were brutal. The two craft used different docking mechanisms entirely, so engineers had to design a common docking adapter that both could mate with.

They breathed different atmospheres: the Apollo cabin held low-pressure pure oxygen, while the Soyuz held a higher-pressure mix closer to ordinary air, which meant a crew member couldn't simply float from one to the other without risking decompression sickness, so the docking module doubled as an airlock to manage the transition. Every interface, mechanical, atmospheric, electrical, communication, language, had to be specified precisely and agreed by both sides, in two languages, across a political divide, with the docking happening once, in orbit, with no chance to fix a mismatch. It worked. The two craft joined, the crews shook hands, and the mission succeeded because both sides had defined and honored a shared, exact specification of every boundary between their systems. That's what an ICD is for, and the Apollo-Soyuz docking remains one of the cleanest examples of why the discipline matters: when you can't iterate, when the parts meet once and have to fit, the written agreement is all that stands between success and an expensive failure in a place you can't reach to repair it.

ICDs feel like bureaucracy on a small project where one person designs both sides of every boundary, and there they may genuinely be overkill. Their value scales with the number of independent parties and the cost of a mismatch. On a system with many teams, many suppliers, and high stakes, the ICD is the mechanism that lets people who never meet build parts that fit, and skipping it doesn't save time, it just defers the cost to integration, where it arrives multiplied.

Design for Reliability

A system that works in the lab but fails in service hasn't solved the problem. **Reliability** is the probability that a system performs its function without failure over a given period under stated conditions, and designing for it means building reliability in from the start rather than hoping to test it in at the end. The deep analysis methods for failure come later in this book; what matters here is the design philosophy, the choices made while the system is being shaped that determine whether it'll hold up.

The first thing to understand is how things actually fail over time, captured in a pattern called the **bathtub curve**. Plot the failure rate of a population of components against their age, and for many systems you get a curve with three regions. Early in life, the failure rate is high, because manufacturing defects and weak units fail quickly; this is called infant mortality. Then the rate drops and stays low and roughly constant for a long stretch, the useful life, where failures happen randomly rather than because of age. Finally, as components wear out, the rate climbs again. The curve's shape explains practices that otherwise seem odd, like burn-in, where manufacturers run components hard for a while before shipping them, deliberately pushing them through the infant-mortality region so the weak units fail in the factory instead of in the customer's hands. It also explains why preventive maintenance and planned replacement make sense: you want to replace wear-out-prone parts before they enter the rising part of the curve, not after.

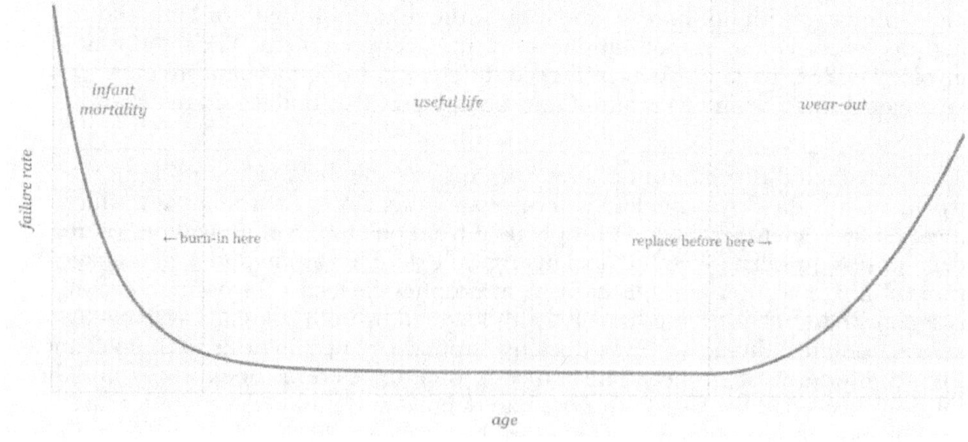

A common way to quantify reliability is **mean time between failures**, or MTBF, the average time a system runs before it fails. A higher MTBF means a more reliable system, and the number lets you compare designs and predict how often you'll see failures across a fleet. But MTBF is easy to misread. A hard drive with a million-hour

MTBF doesn't mean any single drive lasts a million hours, which would be over a century; it means that across a large population during the useful-life portion of the bathtub curve, you'd expect failures at that average rate. Treating MTBF as a guaranteed lifespan for an individual unit is a mistake that's burned plenty of people who didn't read the fine print.

The most important design insight about reliability is also the least intuitive: adding parts usually makes a system less reliable, not more, when those parts are arranged in series. If a system works only when every one of its components works, the components are in series, and the system's reliability is the product of the individual reliabilities. Suppose you have a chain of ten components, each highly reliable at 99 percent. You might expect the system to be roughly 99 percent reliable too. It isn't. Multiply 0.99 by itself ten times and you get about 0.90, so a chain of ten excellent components is only 90 percent reliable as a whole. Stretch it to a hundred such components in series and system reliability falls to about 37 percent, even though every single part is 99 percent reliable. This is why simplicity is a reliability strategy. Every part you add in series is another thing that can fail and take the system down with it, so the fewer parts in the critical chain, the more reliable the system, all else equal. Designers who chase reliability by adding more stuff often make things worse, while designers who achieve it by ruthlessly removing parts from the critical path often succeed.

There are design techniques that improve reliability, and while the deep treatment of redundancy and fault tolerance belongs to the analysis covered later, the design-time choices are worth naming. Simplicity, as just discussed, is the first and most underrated: fewer parts in series means fewer ways to fail.

Derating is a quieter technique with a long track record, running components well below their rated limits so they experience less stress and last longer, like operating a resistor at half its rated power or a capacitor below its rated voltage; the part costs the same but lasts far longer because it's never working hard. Choosing proven components over novel ones improves reliability, because a part with millions of hours of field history is a known quantity, while a brand-new design carries unknown failure modes. And designing for the actual environment, including its extremes, prevents the failures that come from a system meeting conditions its designers didn't account for.

The clearest argument for designing reliability in rather than testing it in is a pair of spacecraft launched in 1977. Voyager 1 and Voyager 2 were built to last about five years, enough to reach Jupiter and Saturn. As of now, both are still operating, still sending data back, nearly fifty years later, from beyond the edge of the solar system, with Voyager 1 the most distant human-made object in existence. They've survived radiation, the cold of deep space, and decades of continuous operation with no possibility of repair, because no one can reach them. That longevity wasn't luck or testing. It was design: generous margins on every component, redundancy in the systems that mattered, conservative derating, proven parts, and a relentless focus on simplicity and robustness during the design phase, when reliability is actually determined. You can't test reliability into a spacecraft after launch, because there are no second chances and no repairs. You have to build it in before it leaves the ground, and the Voyagers are what that looks like when it's done well. The same principle applies to systems you'll never send to deep space: reliability is a property you design in early, not a problem you fix late, and the systems that last are the ones whose designers treated reliability as a requirement from the first sketch.

Reliability done well has reshaped entire industries by changing what's possible. Aircraft engines became so reliable that regulators now let twin-engine airliners fly long routes far from any airport, under rules called ETOPS, which would have been unthinkable when engines failed often enough that you wanted three or four of them and an airport always within gliding distance. The improvement came from decades of designing for reliability: better materials, better understanding of failure modes, derating, and rigorous attention to the things that cause engines to quit. The reward was a fundamental change in how aircraft are designed and operated, all flowing from reliability that was engineered in deliberately over many years.

Design for Manufacturability

A design that can't be built efficiently isn't finished, no matter how clever it is on paper. **Design for manufacturability**, often shortened to DFM, means designing a product so it can actually be made well, at quality, at cost, at the volume required. This sounds obvious, and it's violated constantly, because designers and manufacturers are often different people, and a design that's elegant from a design standpoint can be a nightmare to produce. Bringing manufacturing concerns into the design phase, rather than throwing a finished design over the wall to the factory, is the heart of DFM, and the cost of ignoring it can be ruinous.

The principles of DFM are mostly common sense once stated, though they require discipline to apply. Reduce the number of parts, because every part has to be made, handled, inventoried, and assembled, and fewer parts means lower cost and fewer things to go wrong. Make parts easy to handle and assemble, avoiding designs that require a part to be inserted at an awkward angle or held in place while another part is attached. Use standard components where possible, because custom parts cost more and take longer than off-the-shelf ones. Design parts that can only be assembled the correct way, so an assembler physically can't put them in backward, which eliminates a whole category of defects. Minimize fasteners, because every screw is a part to install, a tool operation, and a chance for error, which is why so much modern product design replaces screws with snap-fits and clips. And design for the manufacturing process you'll actually use, because a part designed for machining looks different from one designed for injection molding or casting, and a design that fights its manufacturing process produces scrap and rework.

There's a formal methodology behind this called **design for manufacture and assembly**, DFMA, developed by Geoffrey Boothroyd and Peter Dewhurst, which gives engineers a structured way to analyze a design and systematically reduce its part count and assembly difficulty. The method asks hard questions about every part: does it need to be separate, or can it be combined with another? Does it move relative to its neighbors? Must it be a different material? If the answer to those questions is no, the part is a candidate for elimination. Companies that have applied DFMA rigorously have cut part counts dramatically, sometimes by more than half, with corresponding drops in cost and assembly time and improvements in reliability, since fewer parts means fewer things to fail. The discipline turns the vague goal of "make it easier to build" into a repeatable analysis with measurable results.

The most expensive recent lesson in what happens when manufacturability is underestimated came from Tesla and its Model 3. In 2017 and 2018, Tesla tried to build a highly automated factory, with robots doing far more of the assembly than was normal in the industry, on the theory that extreme automation would make production faster and cheaper. The plan ran into what Tesla's own leadership later called production hell. The heavily automated line couldn't hit its targets, robots struggled with tasks that humans handle easily, the complex automation broke down and created bottlenecks, and Tesla nearly ran out of cash trying to fix it. The company ended up tearing out some of the automation and bringing back human workers for tasks the robots couldn't manage, and Tesla's leader publicly admitted that the excessive automation had been a mistake and that humans were underrated. The deeper problem was a manufacturing approach that hadn't been matched to what the design and the process could actually support. The vision of a fully automated factory collided with the reality that some assembly steps are genuinely hard to automate, and the design and the production plan weren't reconciled until the company had burned through enormous time and money discovering the mismatch. Manufacturability isn't a detail to sort out after the design is done. It's a constraint that shapes the design, and treating it as an afterthought is how you end up tearing robots out of a factory you just built.

DFM also has to balance against the other design goals, because optimizing purely for manufacturing can hurt the product in other ways. A design that's cheapest to build might be harder to repair, or might compromise performance, or might limit future upgrades. The integrated, glued-together construction that makes some modern electronics cheap and thin to manufacture also makes them nearly impossible to repair, which is a deliberate trade the manufacturer made: easier and cheaper to build, harder for anyone to fix. Whether that trade is right depends on what the product is for and who it's for, and it connects back to the trade-off thinking that runs through all of architecture and design. Manufacturability is one quality among several, and the goal is the right balance for this product, not the maximum on any single dimension.

Model-Based Systems Engineering

For most of the discipline's history, the design of a system lived in documents. Requirements in one set of documents, architecture in another, interfaces in a third, all written in prose and diagrams, stored separately, and updated by hand. The problem with documents is that they drift apart. Someone changes a requirement but doesn't update the interface document that depended on it. The architecture diagram says one thing and the requirements say another, and nobody notices until the contradiction causes a failure. On a large system with thousands of documents maintained by hundreds of people, keeping them all consistent by hand is nearly impossible, and the inconsistencies are where errors hide.

Model-based systems engineering, or MBSE, is the shift away from this scattered pile of documents toward a single connected digital model of the system, where requirements, structure, behavior, and interfaces all live in one linked place and stay consistent because they're parts of the same model rather than separate files.

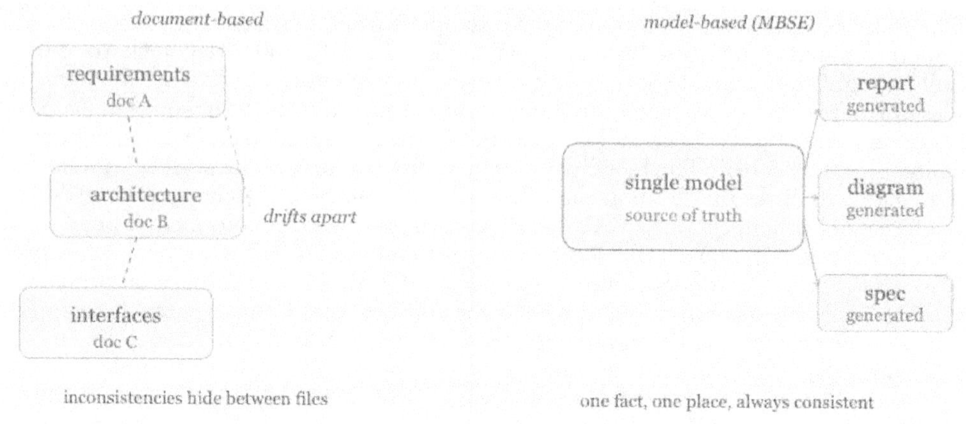

The distinction worth keeping clear is that MBSE is about modeling the description of the system, not simulating its physics. The detailed predictive simulations, the ones that calculate stresses or airflow or temperatures, are a different kind of model covered separately in this book. MBSE's model is descriptive: it captures what the system's parts are, what they do, how they connect, what requirements they satisfy, and how they behave, all as a structured, linked digital representation rather than as prose scattered across documents. The model is the source of truth, and the documents people read, if they still want documents, are generated from the model rather than maintained alongside it.

The language most often used to build these models is **SysML**, the Systems Modeling Language, a standardized notation for representing systems as a set of connected diagrams: diagrams of structure showing how parts relate, diagrams of behavior showing how the system acts over time, diagrams of requirements showing what's required and what satisfies it. Because everything lives in one model, the relationships are explicit and traceable: you can ask the model which components satisfy a given requirement, which requirements a given component touches, and what's affected if you change any one element.

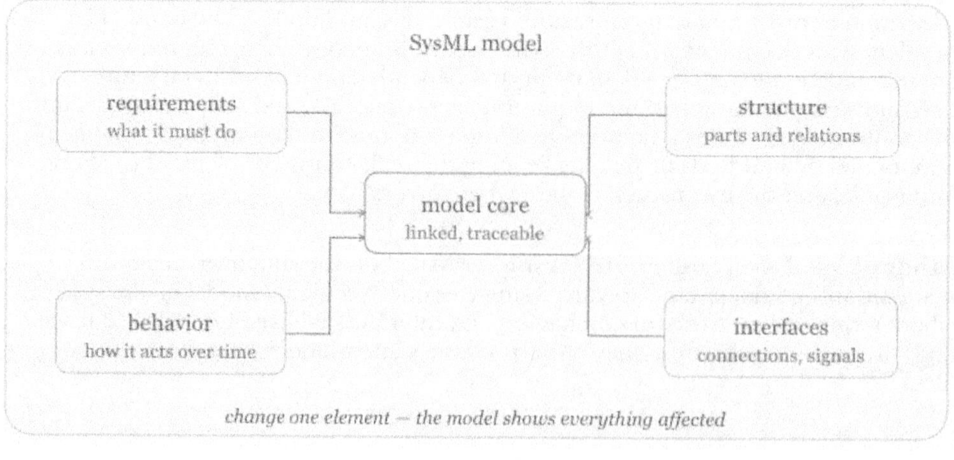

The consistency that's nearly impossible to maintain across documents by hand becomes a property the model enforces, because changing one element updates everything linked to it instead of leaving stale copies scattered around.

The benefits of MBSE are concrete, and they attack the exact failures that document-based engineering produces. Consistency comes first: when everything's in one connected model, the contradictions that hide between documents are far harder to create, because there's one place where each fact lives rather than many copies that can disagree. Traceability comes for free: the links between requirements, design, and verification that have to be maintained painfully in a separate matrix are built into the model's structure. Change impact becomes visible: when you modify something, the model shows you what else is affected, so you update the ripple deliberately instead of discovering the misses later. And communication improves, because a shared model gives everyone the same picture, instead of each team working from its own documents that may have drifted from everyone else's.

Aerospace and defense led the adoption, because their systems are complex enough that document-based engineering had become genuinely unmanageable. NASA's Jet Propulsion Laboratory, which builds and flies interplanetary spacecraft, has been among the leaders in moving to MBSE, using connected models to manage the staggering complexity of missions where a single inconsistency can lose a spacecraft and there's no chance to repair it in flight. Major aircraft manufacturers and defense programs have followed, because a modern aircraft or weapons system has so many interconnected requirements, components, and interfaces that keeping them consistent in prose documents had become a losing battle. The move to MBSE in these industries isn't a fashion; it's a response to systems that had outgrown the document-based methods used to design them.

MBSE isn't free, and the honest account includes its costs. Building and maintaining a model takes skill and tooling that not every organization has, the modeling languages have a learning curve, and the discipline of keeping the model current requires real commitment, because a model that's allowed to fall out of date is worse than no model, since people trust it and it lies to them. On a small, simple system, the overhead of formal modeling may exceed the benefit, and documents may genuinely be the right choice. The value of MBSE rises with complexity, the same way the value of interface control documents and traceability does. For a small project, it may be more than you need. For a system complex enough that no one can hold it in their head and many teams have to stay consistent with each other, MBSE is increasingly how serious organizations keep the system coherent, because the alternative, a growing pile of documents drifting quietly out of agreement, fails in exactly the ways the discipline exists to prevent.

Discussion question: If every architectural decision is a trade-off with no universally correct answer, and if the right choice depends entirely on which qualities matter most for a particular system, then how can an architect justify a decision to stakeholders who each value different qualities, and what makes an architectural choice defensible rather than merely a matter of the architect's personal preference?

Chapter 4: Modeling and Simulation

Before a single Boeing 777 ever flew, the airplane existed. It existed as a complete digital model, every one of its roughly three million parts represented in software, fitted together and checked for clashes inside a computer. Boeing built the 777 in the early 1990s as the first jetliner designed entirely without a physical mockup, the full-size wood-and-metal model that aircraft makers had always built to check whether the parts actually fit. Instead of building the plane to find the problems, Boeing found the problems in the model and built the plane right the first time. The approach cut the costly errors that used to surface during assembly, when fixing them meant tearing apart real hardware. The airplane was designed, in a real sense, twice: once as a model and once as metal, and the model came first so the metal would come out right.

That's the promise of modeling and simulation, and it's why the discipline has become central to building anything complex. A model lets you ask "what happens if?" without paying the full price of finding out for real. You can crash a car a thousand times in simulation before crashing one on a test track. You can run a factory for a year of simulated time in an afternoon. You can fly a spacecraft to Mars in software and discover the trajectory error before it's burned into an actual mission. This chapter covers the main kinds of models systems engineers use: physics-based models that compute how the physical world behaves, discrete event simulations that capture systems where things happen at distinct moments, digital twins that mirror a real operating system in real time, and the question that hangs over all of them, which is how much you can trust a model before you bet something real on its answer.

Physics-Based Models

A **physics-based model** computes how a system behaves by applying the actual laws of physics: the equations governing how forces act, how heat flows, how fluids move, how structures bend and break. Rather than guessing or relying on rough rules of thumb, these models solve the underlying physics to predict what a real system will do under conditions you specify. When you see a simulation of air flowing over a wing, stress spreading through a bridge under load, or heat building up inside an engine, you're looking at a physics-based model doing the math that physical reality does on its own.

The workhorse technique behind most of these models is the **finite element method**, and the core idea is approachable even though the math behind it is heavy. Real physical objects are continuous, and the equations describing how they behave can't usually be solved exactly for a complicated shape. So the finite element method breaks the object into a large number of small, simple pieces called elements, a mesh of tiny shapes covering the whole structure, and solves the physics approximately within each small piece, then stitches the pieces together. Each element is simple enough to handle, and the assembly of thousands or millions of them approximates the behavior of the whole continuous object. The finer the mesh, the more accurate the result and the more computing it takes, which is a trade-off that runs through all of physics-based modeling: accuracy costs computation, and you're always choosing how much of each you can afford.

A related family of techniques handles fluids. **Computational fluid dynamics**, or CFD, models how gases and liquids move: air over an aircraft, water around a ship's hull, blood through an artery, coolant through a reactor. Fluids are harder to model than solid structures because they swirl, mix, and turn turbulent in ways that are genuinely difficult to compute, and turbulence in particular remains one of the hard unsolved problems of physics, approximated rather than solved exactly even in the best models. CFD is what lets engineers shape a car for low drag, design a quieter aircraft, or predict how smoke will spread through a burning building, all without building the thing and measuring it.

The reason physics-based modeling matters so much is the brutal economics of the alternative. Physical testing of large or dangerous or expensive systems is slow, costly, and sometimes impossible. You can't crash-test a skyscraper. You can't easily run a hundred variations of a jet engine to find the best one when each engine costs millions and takes months to build. A model lets you explore the design space cheaply, trying configurations that would be impractical to test physically, and narrowing down to a few good candidates before you commit to building anything. The savings are enormous, in money and in time, and in some cases the model makes possible things that couldn't be done any other way, because the physical experiment is simply out of reach.

Crash safety is where physics-based modeling has saved the most lives, and the change over a few decades is dramatic. Carmakers once developed crash safety almost entirely by building cars and smashing them into barriers, an expensive process that destroyed a car with every test and yielded only as many data points as you could afford to wreck. Now most of that work happens in simulation. Engineers model the entire car and a crash in software, computing how the structure crumples, how the energy of the impact moves through the frame, how the occupants and airbags respond, all before any physical car is built. They can run thousands of simulated crashes, varying the speed, the angle, the obstacle, the size of the occupant, exploring far more scenarios than physical testing could ever reach, and using physical crash tests mainly to confirm what the models predicted.

The result is cars that protect people in a wide range of crashes the designers anticipated through simulation, and a development process that finds and fixes safety problems in the model, where fixing them is cheap, rather than in physical tests or, worse, in real crashes on the road. Modern vehicle safety, the kind that lets people walk away from collisions that would have been fatal a generation ago, rests heavily on physics-based models that let engineers understand and improve crash behavior without destroying a car to learn each lesson.

These models have limits that matter, and the most important one is captured in a line statisticians repeat: all models are wrong, but some are useful. A model is always a simplification of reality, leaving things out, making assumptions, approximating what it can't compute exactly. It captures the physics the modeler chose to include and ignores the rest, and the parts left out are sometimes the parts that turn out to matter. A structural model might assume a material behaves uniformly when the real material has flaws. A fluid model might approximate turbulence in a way that's close enough most of the time and badly wrong in some unusual condition. The model's answer is only as good as its assumptions, and a model that's accurate across the range of conditions it was built for can be confidently, dangerously wrong outside that range. This is why the question of how much to trust a model is so serious, and why it gets its own treatment later in this chapter. A physics-based model is one of the most useful things in engineering, and it's also a simplification that can mislead you precisely because it looks so authoritative, producing crisp numbers and colorful pictures that carry an air of certainty the underlying assumptions don't always deserve.

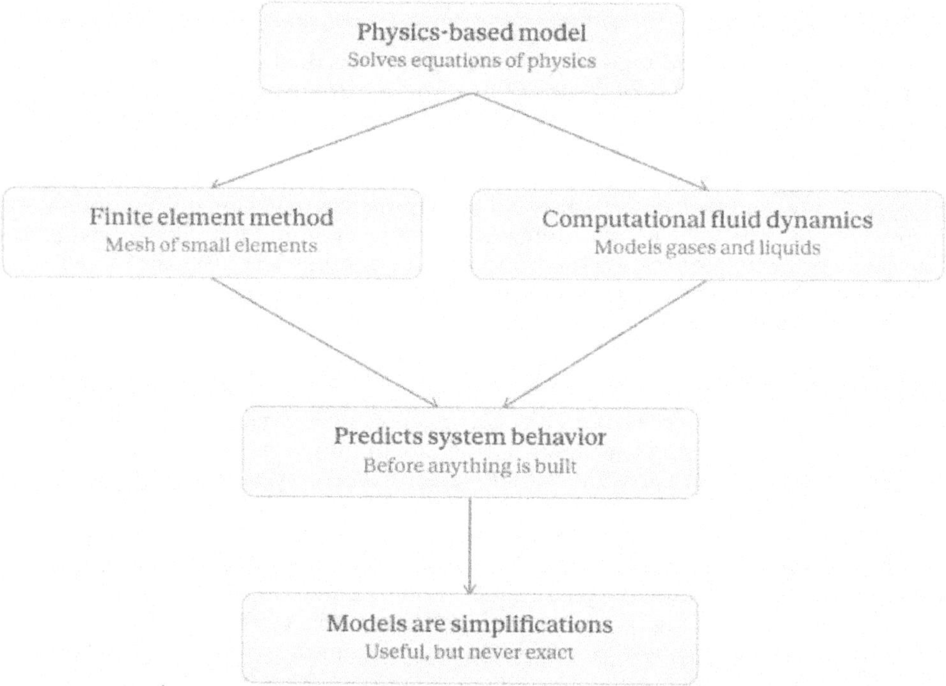

Discrete Event Simulation

Not every system is best understood through continuous physics. Many systems are about things happening at distinct moments: a customer arrives, a part finishes processing, a truck departs, a server completes a request. For these, the relevant model is **discrete event simulation**, which models a system as a sequence of separate events occurring at specific points in time, jumping from one event to the next rather than computing continuous change. Between events, nothing the model cares about is happening, so the simulation simply skips ahead to the next event, which makes it possible to simulate long stretches of real time very quickly.

The kinds of systems this suits are everywhere in operations. A factory floor, where parts move between machines, wait in queues, and get processed, is a sequence of discrete events: this machine starts, that one finishes, a part joins a queue, a worker becomes available. A hospital emergency room, where patients arrive unpredictably, wait, get seen, occupy beds, and leave, is the same kind of system. An airport, a call center, a shipping port, a supply chain, a checkout line at a store, are all systems where discrete things happen at discrete times and the questions you care about are about flow, waiting, and capacity rather than forces and temperatures.

The concepts that show up repeatedly in these models are worth naming, because they're how you think about any system of flow and queuing. There are entities, the things that move through the system, like customers or parts or patients. There are resources, the things that serve them, like machines, workers, or beds, which are limited and get occupied and freed. There are queues, where entities wait when the resource they need is busy. And there's randomness, because real arrivals and service times aren't fixed, they vary, and the variation is often the whole story. A model that assumed every customer arrived exactly five minutes apart and every service took exactly four minutes would show no waiting at all. Real systems have waiting precisely because the timing is random, with clumps and gaps, and discrete event simulation captures that randomness, which is what makes its answers realistic where simple average-based calculations fail.

The reason these models are so useful is that systems of queues behave in ways that defy intuition, and the most important counterintuitive fact is about what happens as a system gets busy. You might expect that as utilization climbs toward a hundred percent, waiting times rise steadily, doubling as you go from comfortable to packed. They don't. Waiting time rises gently while utilization is moderate, then climbs sharply as utilization approaches full capacity, and near a hundred percent it shoots toward infinity. A system running at 70 percent capacity might have short, manageable queues, while the same system at 95 percent has waiting times many times longer, and at 99 percent the queues explode.

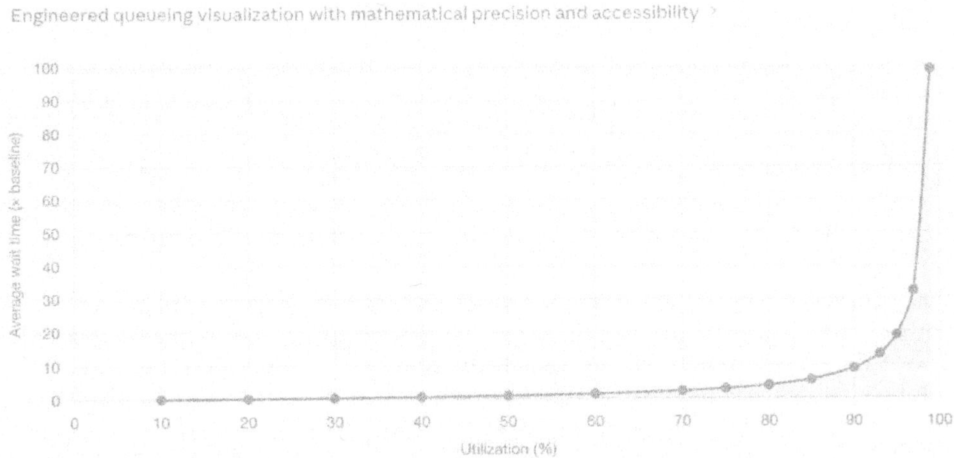

This is why a hospital that runs its beds at very high occupancy to save money ends up with patients waiting in hallways: the high utilization that looks efficient on a spreadsheet produces enormous waiting in reality, because the system has almost no slack to absorb the random clumps of arrivals. Discrete event simulation reveals this behavior, lets you find the utilization level where waiting stays acceptable, and prevents the costly mistake of running a system so "efficiently" that it chokes on its own randomness.

Amazon's warehouses show what discrete event simulation makes possible at scale. A modern fulfillment center is a staggeringly complex system of flow: items arrive, get stored, get retrieved when ordered, move to packing, get boxed, and ship out, with robots and workers and conveyors all moving things through a sequence of steps, millions of times a day. Designing such a facility, deciding how many robots, how many packing stations, how to lay out the storage, how to route items, is impossible to optimize by intuition, because the interactions between all those moving parts produce flow behavior no one can predict by thinking about it. So these systems are modeled and simulated before they're built and continuously as they run: engineers simulate the facility, try different numbers of resources and different layouts and different routing rules, watch where queues form and where bottlenecks appear, and tune the design in the model before committing to the physical buildout. The choreography that lets a vast warehouse fill orders quickly without items piling up at a bottleneck is worked out in simulation, where a bad layout costs nothing but computing time, rather than in concrete, where it would cost a building.

The value of discrete event simulation is highest exactly where intuition fails hardest, which is in systems with many interacting parts, significant randomness, and queues that build in nonlinear ways. For a simple system you can reason about directly, the simulation may be unnecessary. For a complex operation where the flow behavior emerges from countless interactions and the cost of getting the capacity wrong is high, the simulation is often the only way to understand the system before you build it, and the only way to find the configuration that works without the painful method of building the wrong thing and discovering its problems in operation.

Digital Twins

The models discussed so far are built before a system exists or runs separately from it. A **digital twin** is different: it's a living digital model of a specific real system that exists now and is operating, kept continuously in sync with its physical counterpart through a stream of data from sensors on the real thing. The physical system and its digital twin run in parallel, the twin updated in near real time by data flowing from the real system, so the model always reflects the current actual state of that specific physical object, not a generic design but this exact machine as it is right now.

The distinction that makes a digital twin special is the continuous, two-way connection to a particular physical thing. A regular model represents a design or a type. A digital twin represents one specific instance, this turbine with this serial number, fed by sensors on that turbine, reflecting its real current condition including all the wear and the quirks and the history that make it different from an identical unit next to it. Data flows from the physical system to the twin to keep the model current, and insights flow back from the twin to the physical system to guide how it's operated and maintained. The twin is a digital shadow that tracks its physical original through the original's actual life.

What this enables is a set of things you can't do with a model that isn't connected to reality. You can monitor the real system's condition through the twin, seeing its current state in detail even where you can't easily measure the physical thing directly.

You can predict failures before they happen, because the twin, fed by real data, can show stress accumulating or a component degrading, letting you intervene before it breaks rather than after. You can test changes safely, trying an adjustment on the twin first to see what it does before making it on the real system, which matters enormously when the real system is expensive or dangerous to experiment on. And you can optimize operation continuously, using the twin to find better ways to run the actual system as conditions change, rather than relying on settings chosen once at design time for conditions that may no longer hold.

General Electric built much of its modern industrial business around digital twins of jet engines and turbines, and the application shows the concept's value clearly. A jet engine is enormously expensive, operates under brutal conditions, and absolutely cannot fail in flight, which makes knowing its true condition both vital and difficult. GE equips its engines with sensors that stream data, temperatures, pressures, vibrations, rotational speeds, throughout operation, and feeds that data into a digital twin of each specific engine.

The twin reflects the actual condition of that individual engine as it accumulates hours and wear, which lets GE predict when a particular engine will need maintenance based on its real condition rather than on a fixed schedule that ignores how hard this specific engine has actually been worked. An engine that's flown gentle routes in clean air ages differently from one that's flown through dust and heat, and the digital twin captures that difference, so maintenance happens when this engine actually needs it. The payoff is engines that are serviced before they develop problems, that aren't pulled from service for maintenance they don't yet need, and whose true condition is known continuously rather than guessed at between inspections. The same approach now runs across wind turbines, power plants, and heavy industrial equipment, anywhere the cost of a failure is high and the value of knowing a machine's real condition justifies the sensors and the modeling to maintain a twin.

Digital twins depend on things that aren't always present, and the honest view recognizes the requirements. A twin needs a flow of sensor data from the real system, which means the physical thing has to be instrumented, and instrumenting it costs money and isn't always feasible. It needs a model good enough that the twin actually reflects reality, which brings back the whole question of model fidelity. And it needs the computing and the connectivity to keep the twin current. For a high-value, sensor-rich system like a jet engine or a power plant, all of this is justified by what the twin saves. For a cheap or simple system, the cost of building and maintaining a twin may far exceed the benefit, and a digital twin would be elaborate overkill. The concept is genuinely useful, and it's also frequently oversold, attached to systems that don't justify it because the term has become fashionable. The judgment is the same one that runs through this whole book: the value of the technique has to be weighed against its cost for the specific system at hand, and a digital twin earns its place only where the stakes and the data justify the effort to build and maintain it.

Verification of Simulation Fidelity

Here is the question that should haunt anyone who relies on a model: how do you know the model is right? A model produces confident, precise answers, crisp numbers and vivid pictures that look authoritative, and it can be completely wrong.

Fidelity is how accurately a model represents the real system, and **verification of simulation fidelity** is the work of establishing how much you can actually trust a model before you bet something real on its predictions. This is the discipline that keeps modeling honest, and skipping it is how a model becomes a way to be confidently wrong at scale.

Two distinct questions hide inside "is the model right?", and keeping them separate matters because they're checked differently.

Verification asks: did we build the model correctly? Does the software actually compute what we intended, free of bugs, solving the equations the way we meant it to? A model can be a faithful implementation of an idea that's wrong about reality, so verification only confirms that the model does what its builders intended, not that what they intended matches the world.

Validation asks the harder question: did we build the right model? Does the model actually match reality? You answer validation by comparing the model's predictions against measurements from the real system, and a model can pass verification, being a clean and bug-free implementation, while failing validation completely, because the physics it implements doesn't capture what really happens. The pairing is sometimes summarized as verification checking whether you built the thing right, and validation checking whether you built the right thing, and a trustworthy model needs both.

The deepest danger in modeling isn't a model that's obviously broken, because an obviously broken model gets caught. The danger is a model that's plausible and wrong, one that produces reasonable-looking answers that happen to be incorrect, because those are the answers people believe and act on. A model that crashes or returns nonsense announces its own failure. A model that returns a confident, sensible-looking number that's wrong by a factor that matters gives no warning at all, and its very plausibility is what makes it dangerous, because people trust it precisely because it looks right. This is why validation against real data isn't optional for any model whose answers carry consequences: the model's plausibility is not evidence of its correctness, and the only way to know whether a model matches reality is to check it against reality.

The catastrophe that should be taught alongside every modeling course is the loss of the **space shuttle Columbia** in 2003. During launch, a piece of insulating foam broke off the external tank and struck the leading edge of the shuttle's wing. The question of whether this damage was dangerous was assessed in part using a model, a piece of software called Crater that predicted the damage from debris impacts. Crater predicted the foam strike would cause only shallow damage, not a threat to the vehicle. But Crater had been built and calibrated for impacts from small debris, and the piece of foam that hit Columbia was far larger than anything in the model's validated range, hundreds of times the size of the debris the model was designed around. The model was being used far outside the conditions for which it had been validated, where its predictions had no established reliability, and it predicted survivable damage for an impact that in reality had punched a hole in the wing's thermal protection. On reentry, hot gas entered through that breach and destroyed the wing, and the shuttle broke apart, killing all seven astronauts aboard. The investigation found many causes, organizational and technical, but the modeling lesson is stark and specific: a model trusted outside its validated range gave a confident answer that was lethally wrong, and the people relying on it didn't adequately account for the fact that they were asking the model a question it had never been shown to answer correctly. The model wasn't lying. It was being used for something it was never validated to do, and its confident output concealed that it had no business making that prediction.

That tragedy points to the practices that keep model use honest, and they're worth stating as discipline rather than suggestion. Know the validated range of any model, the conditions under which it's been checked against reality, and treat its predictions outside that range as unproven, however confident they look. Quantify the uncertainty rather than treating a model's output as an exact truth, because every model has error, and knowing roughly how much error helps you judge how much to trust the answer and how much margin to leave. Compare against real data whenever you can, because validation against reality is the only genuine test of whether a model matches the world. And keep a healthy skepticism toward results that look too clean, since reality is messy and a model that produces suspiciously tidy answers may be hiding its assumptions rather than capturing the truth. Trust has to be earned through validation and bounded by knowing where the validation stops, rather than granted automatically because the output looks authoritative.

There's a cultural dimension to this that the Columbia case exposed and that matters in every organization that uses models. Models carry an authority that can suppress the human judgment that should check them. A confident simulation result can shut down the engineer who has a bad feeling about it, because who argues with the computer and its precise numbers and colorful graphics? But the engineer's unease is sometimes exactly the signal that the model is being trusted beyond its range, and an organization that lets the model's apparent authority override human concern has set itself up for the kind of failure that destroyed Columbia.

The model is a tool that informs judgment, not a replacement for it, and the people using it have to retain both the knowledge to understand its limits and the standing to question its answers when something seems wrong. A model that can't be questioned isn't a source of truth. It's a source of unchallenged error waiting for the conditions that expose it, and the discipline of verifying fidelity is, in the end, about keeping human judgment in charge of the machine rather than the other way around.

Discussion question: If a model can only be validated within the range of conditions where you have real data to check it against, but the whole reason you build a model is often to explore conditions you can't or haven't tested in reality, then how can a model ever be trusted for the novel situations it's most needed for, and what should an engineer do when a decision depends on a prediction that lies beyond anything the model has been proven to get right?

Chapter 5: Risk and Reliability Analysis

In April 1970, an oxygen tank exploded aboard Apollo 13 on its way to the Moon, crippling the spacecraft and turning a lunar landing into a struggle to get three astronauts home alive. The explosion came from a chain of small things nobody had connected: a tank that had been dropped years earlier, a heater thermostat rated for the wrong voltage, ground crews who'd boiled off the tank's contents during a test using a procedure that quietly damaged the wiring inside. No single mistake would have caused the disaster. The failure emerged from the combination, and the combination was exactly the kind of thing that careful risk analysis exists to find before it kills someone, rather than after. The crew survived through extraordinary improvisation, but the lesson NASA took was about what should have been caught on the ground: the failure modes nobody had traced, the single points whose loss was catastrophic, the small defects that lined up into a chain.

This chapter is about finding those failures before they happen. Every system can fail, and the engineer's job isn't to pretend otherwise but to understand how a system fails, how likely each failure is, and what to do about the ones that matter. The methods here are the formal tools of that work: failure mode and effects analysis, which walks through every way each part can fail; fault tree analysis, which works backward from a disaster to its causes; probabilistic risk assessment, which puts numbers on the likelihood of failure; redundancy and fault tolerance, which keep systems working when parts break; human factors, because people are part of every system and a major source of its failures; and margin and robustness, the discipline of building systems that survive conditions worse than expected. Together these methods turn "I hope it doesn't break" into a structured understanding of exactly how it might break and what stands between a failure and a catastrophe.

One of the uncomfortable realities of risk analysis is that the most dangerous failures are often not the most likely ones. Engineers can spend enormous effort preventing common, low-consequence problems while overlooking a rare event whose probability seems too small to worry about until it happens. The challenge isn't simply identifying what can go wrong, but balancing likelihood against consequence. A failure that occurs once every million operating hours may deserve little attention if it causes a brief inconvenience, while a failure with the same probability that could cause fatalities or destroy a billion-dollar asset demands intense scrutiny.

Risk analysis is ultimately an exercise in prioritization, deciding which risks can be accepted, which must be reduced, and which are so severe that the system itself must be redesigned. The goal is not absolute safety, which is impossible, but an informed understanding of where danger lives and how much uncertainty remains.

Failure Mode and Effects Analysis

The most widely used method for thinking systematically about failure is **failure mode and effects analysis**, almost always called FMEA. The method is conceptually simple and relentlessly thorough: you go through a system part by part, and for each part you ask how it can fail, what happens when it does, and what that means for the whole system. Done completely, FMEA forces you to confront every way each component can break and trace the consequences, which surfaces dangers that would otherwise stay hidden until they bit you in operation.

The vocabulary matters because the method depends on distinguishing three things.

A **failure mode** is the way a component fails: a valve can fail stuck open, fail stuck closed, or fail leaking, and each is a distinct mode.

The **failure effect** is what happens as a result: if the valve fails stuck closed, fluid stops flowing, and downstream the system loses pressure, and beyond that something it powers shuts down.

The **failure cause** is why the failure happens: corrosion, wear, a manufacturing defect, contamination. A single component can have several failure modes, each with its own effects and causes, and FMEA insists on considering all of them rather than just the one that comes to mind first, because the failure mode you don't think about is the one that gets you.

The method becomes a priority tool through a number called the **risk priority number**, or RPN, which combines three judgments about each failure mode. Severity rates how bad the consequences are if the failure happens, usually on a scale of one to ten, where a minor annoyance scores low and a catastrophic, life-threatening effect scores high. Occurrence rates how likely the failure is, where something that almost never happens scores low and something frequent scores high. Detection rates how likely you are to catch the failure before it causes harm, and here the scale runs backward in a way that trips people up: a failure you'll easily detect scores low, because easy detection is good, while a failure that's nearly impossible to catch before it does damage scores high, because hidden failures are dangerous. Multiply the three together, severity times occurrence times detection, and you get the RPN, a single number that ranks failure modes by how much attention they deserve. A failure that's severe, likely, and hard to detect produces a high RPN and demands action. One that's minor, rare, and easy to catch produces a low RPN and can wait.

The detection factor is the one worth dwelling on, because it captures something people get wrong about risk. A severe failure isn't automatically the most dangerous one. A severe failure that you'll always detect before it causes harm is far less dangerous than a moderate failure that strikes without warning, because the warning is what lets you act. A brake system that fails gradually and lights a warning lamp gives the driver time to respond; a brake system that fails suddenly and silently doesn't. This is why detection sits in the calculation alongside severity and likelihood: the failures that hurt people most are often not the most severe in raw terms but the ones that arrive with no warning, where detection is poor, and FMEA's structure forces you to weigh that explicitly rather than fixating on severity alone.

The automotive industry runs on FMEA, and the reason is the scale of the consequences when a car fails. A modern vehicle has thousands of components, any of which can fail in several ways, and a failure in the field means a car that breaks down, or crashes, or gets recalled by the millions at staggering cost. So carmakers apply FMEA throughout design, working through the failure modes of every component and subsystem, ranking them by RPN, and attacking the high-priority ones before the car is built. The discipline is built into industry standards that suppliers must follow, because the cost of a failure mode that escapes into production is so high. When a part is found during design to have a high RPN, the engineers change something: they make the part more reliable to lower its occurrence, or add a way to detect its failure to lower its detection score, or redesign the system so the failure's effect is less severe. The goal is to drive down the RPNs of the dangerous failure modes before any physical car exists, because a failure mode caught in an FMEA costs a design change, while the same failure mode caught in the field costs a recall and possibly lives.

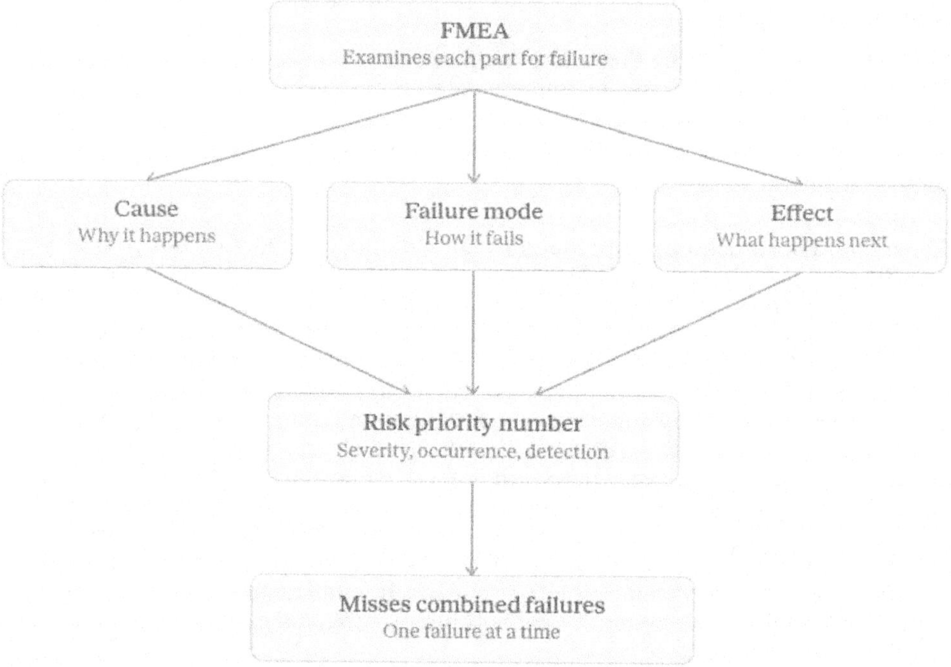

FMEA has a structural blind spot that you have to understand, because relying on it alone gives a false sense of safety. The method works one failure at a time: it asks what happens when this part fails, assuming everything else works. That makes it excellent at catching single-point failures, the cases where one component's failure causes trouble on its own. But it's poor at catching combinations, the failures that only happen when two or more things go wrong together, because FMEA's part-by-part structure doesn't naturally consider multiple simultaneous failures. Apollo 13's explosion came from a combination, several things going wrong in sequence, which is exactly the pattern FMEA struggles to see. The real world produces these combination failures regularly, where each individual fault is survivable but their coincidence is catastrophic, and a team that has done a thorough FMEA and concluded the system is safe may have missed entire categories of multi-failure scenarios. This is precisely why the next method exists, because the limitation of working forward one part at a time creates the need for a method that works backward from the disaster and can capture how combinations of failures produce it.

Fault Tree Analysis

Where FMEA starts from components and works forward to effects, **fault tree analysis** starts from a disaster and works backward to its causes. You name a specific catastrophic event you want to prevent, the **top event**, and then you build a tree of all the things that could cause it, branching downward through intermediate causes to the basic faults at the bottom. The method is top-down where FMEA is bottom-up, and that reversal is exactly what lets it capture the combination failures that FMEA misses.

The tree is built using logic gates that capture how causes combine, and two gates do most of the work. An **OR gate** means the event above it happens if any one of the events below it happens: a car won't start if the battery is dead, or the starter is broken, or the fuel is empty, any single one is enough. An **AND gate** means the event above happens only if all of the events below happen together: a room goes dark only if the main power fails and the backup generator also fails and the emergency lights also fail, all at once. The distinction is the heart of the method, because it's how the tree represents the difference between a single failure that's enough on its own and a combination of failures that all have to occur together. FMEA, working one part at a time, can't easily express "this and this and this together," but a fault tree expresses it directly through AND gates, which is why fault trees find the multi-failure scenarios that part-by-part analysis overlooks.

The most valuable thing a fault tree reveals is the **single point of failure**, a single fault that's enough, by itself, to cause the top event. In the tree, a single point of failure shows up as a basic fault that leads to the top event through nothing but OR gates, with no AND gate requiring anything else to also fail. These are the most dangerous features of any system, because they mean one component's failure brings down the whole thing with no backup to catch it, and finding and eliminating them is one of the main reasons to build a fault tree. When the tree exposes a single point of failure for a catastrophic event, the engineers know they need to add redundancy or a safeguard so that no single fault can cause the disaster alone, converting that part of the tree from a bare OR path into something that requires multiple failures through an AND gate.

The analysis goes further by finding what are called **cut sets**, the combinations of basic faults that together cause the top event. A minimal cut set is the smallest such combination, the fewest things that have to fail together to produce the disaster, and the size of the smallest cut sets tells you how vulnerable the system is. If the smallest cut set has just one fault in it, you have a single point of failure and the system is fragile, because one thing breaking is enough. If the smallest cut sets all contain three or four faults, the system is robust against catastrophe, because the disaster requires several independent things to fail at once, which is far less likely. Mapping the cut sets shows you exactly which combinations to worry about and lets you compare designs by how many simultaneous failures it takes to defeat them.

Nuclear power plants are where fault tree analysis was developed and where it's applied most rigorously, for the obvious reason that the top events you're trying to prevent, a core meltdown, a release of radiation, are catastrophic enough to justify any amount of analysis. The methodology grew out of the nuclear and aerospace industries in the 1960s precisely because these were the first systems where a major failure was unthinkable and the cost of analysis was trivial next to the cost of the disaster. A nuclear plant's safety case is built around fault trees for its catastrophic events, tracing every path to meltdown, finding every single point of failure and eliminating it, identifying the cut sets and ensuring that no small combination of failures can cause a release. The plants are then designed so that catastrophe requires many independent failures together, the deep safety that comes from a system where the minimal cut sets are large.

This is why nuclear accidents are rare, and why the ones that happen, like the chain of failures at Fukushima in 2011, tend to involve exactly the kind of multi-failure scenarios that defeat a system's redundancy, often through a common cause that fails several supposedly independent safeguards at once. At Fukushima, the earthquake and then the tsunami knocked out both the main power and the backup generators together, a common cause that defeated the redundancy because the backups weren't independent of the threat, sitting where the same wave that caused the emergency also destroyed the systems meant to handle it.

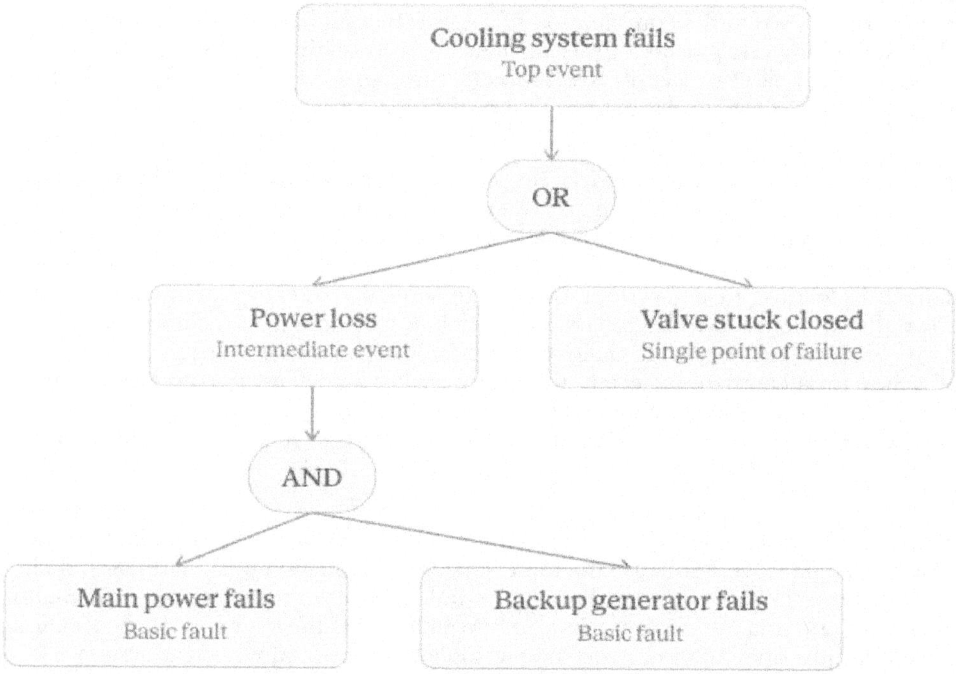

(The right branch is the dangerous one. A stuck valve alone takes down cooling, no other failure required, so it's a single point of failure. The left branch needs main power and the backup generator to fail at the same time, which is what an AND gate buys you: the same outcome, but only if two independent things go wrong together instead of one.

This is also the shape of the Fukushima problem. The tree treats main power and backup generator as independent, but if one event (the tsunami) can knock out both at once, the AND gate's protection disappears and the left branch behaves like the right one.)

That common-cause problem is the subtle limit of fault tree analysis and of redundancy generally, and it deserves to be flagged here because it undermines the comfort that large cut sets provide. The analysis assumes the basic faults are independent, that the probability of several failing together is the product of their individual probabilities, which is very small. But if a single underlying cause can trigger several faults at once, a flood, a fire, a design error common to redundant units, a software bug present in every copy, then those faults aren't independent, and the combination that the tree treated as nearly impossible becomes a single event. A system with triple redundancy looks robust because all three units must fail, an unlikely coincidence if they fail independently, but if all three share a power supply, or sit in the same room, or run the same flawed software, then one common cause takes out all three together, and the redundancy was an illusion.

Good fault tree analysis hunts specifically for these common causes, because they're how systems that look safe on paper fail in reality, and the discipline of asking "what single thing could defeat all my supposedly independent safeguards at once" is one of the most valuable habits the method instills.

Probabilistic Risk Assessment

FMEA and fault tree analysis identify how things fail. **Probabilistic risk assessment**, or PRA, puts numbers on it, combining the likelihood of failures with the severity of their consequences to produce a quantified picture of risk. Rather than saying a failure is "unlikely" or "serious," PRA tries to say how unlikely and how serious in numbers, so that risks can be compared, ranked, and weighed against the cost of reducing them. The basic formula is straightforward: risk equals the probability of an event times the consequence of that event, so a rare event with catastrophic consequences and a frequent event with mild consequences can be placed on the same scale and compared directly.

The reason to quantify risk rather than just describe it is that resources are finite and intuition about probability is terrible. You can't eliminate every risk, so you have to choose which ones to spend money and effort on, and that choice should go to the risks that contribute most to the total, which means the ones with the worst combination of likelihood and consequence. Without numbers, people prioritize by how frightening a risk feels rather than how much it actually contributes, and human fear is a poor guide to actual risk: people overestimate dramatic, memorable dangers and underestimate quiet, common ones. PRA cuts through this by forcing the comparison onto a numerical footing, where a risk that feels terrifying but contributes little to the total can be correctly deprioritized in favor of a risk that feels mundane but dominates the numbers.

The honest difficulty with PRA is that the numbers are often deeply uncertain, and pretending otherwise is dangerous. For some failures you have good data: components that fail often enough, across large populations, that you can measure their failure rates with confidence. For others you have almost nothing: rare events that have happened a handful of times or never, novel designs with no track record, human errors whose probability is genuinely hard to pin down.

The probability of a common component failing might be known to within a few percent, while the probability of an unprecedented catastrophe might be uncertain by a factor of ten or a hundred, and a PRA that presents both with the same crisp decimal places hides that difference in a way that misleads. The output of a PRA is only as good as the probabilities fed into it, and the probabilities for the rarest and most consequential events are exactly the ones hardest to know, which means the assessment is often most uncertain precisely where the stakes are highest.

The danger in PRA isn't the math, which is sound. It's false precision, the way putting a specific number on a deeply uncertain risk can create confidence that the number doesn't deserve. When a PRA reports that the probability of a catastrophic failure is one in a million per year, that number looks authoritative, and people treat it as established fact, when in reality it may be a rough estimate built on assumptions that could be wrong by orders of magnitude. The precision of the output conceals the uncertainty of the inputs, and a decision made by trusting the number as if it were measured rather than estimated can be badly wrong. This is the same plausibility trap that afflicts simulation models: a confident, precise number that's actually a guess is more dangerous than an honest "we don't know," because the false precision suppresses the caution that genuine uncertainty would invite.

NASA learned about misplaced confidence in probability numbers in the most public way possible, and the physicist Richard Feynman made the lesson permanent. After the space shuttle Challenger exploded in 1986, killing all seven crew, Feynman served on the investigation and probed how NASA estimated the risk of the shuttle failing. He found a startling gap. NASA management was using a failure probability of roughly one in a hundred thousand per flight, an extraordinarily low number implying you could fly the shuttle every day for centuries and expect no losses. The engineers who actually built and worked on the hardware estimated the risk closer to one in a hundred, a thousand times higher. Feynman concluded that the management number wasn't derived from any honest analysis of the hardware; it was a number that reflected what management wanted or needed to believe rather than what the engineering supported. His report ended with a line that's quoted in every serious discussion of risk: for a successful technology, reality must take precedence over public relations, because nature cannot be fooled. The shuttle's true risk was roughly what the engineers said, as the loss of Challenger and later Columbia, two losses in 135 flights, grimly confirmed, landing the real failure rate close to the engineers' estimate and nowhere near management's. The lesson is that a risk number is only as honest as the analysis behind it, and a number chosen to reassure rather than to reflect reality is worse than useless, because it justifies decisions that the true risk would forbid.

Used honestly, PRA is genuinely valuable, and the honest use has a particular character. It compares risks so resources go where they matter most. It exposes which failures dominate the total risk, often a small number of contributors that account for most of the danger, so effort concentrates there. And, used with integrity, it presents its uncertainties openly rather than hiding them behind false precision, reporting not just a number but a range, and being honest about which inputs are well known and which are guesses. The discipline isn't in the arithmetic, which any competent analyst can do. It's in the intellectual honesty about how much the numbers actually mean, the willingness to say "this estimate could be off by a factor of fifty" when that's true, and the refusal to let a precise-looking output substitute for genuine understanding of a risk that remains, underneath the decimals, uncertain.

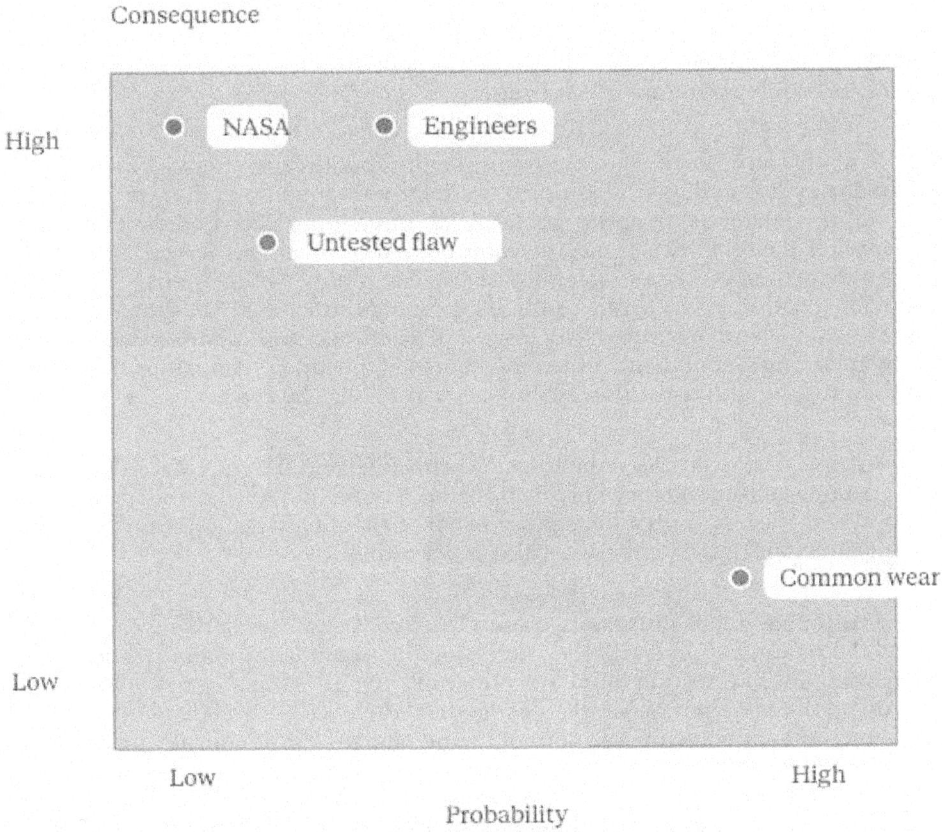

Redundancy and Fault Tolerance

The analysis methods tell you where a system can fail. **Fault tolerance** is the design property that lets a system keep working even when parts of it fail, and **redundancy**, having more than one way to accomplish something so that a backup takes over when the primary fails, is the main technique for achieving it. A fault-tolerant system is one designed on the assumption that things will break, with the means to survive those breakages built in rather than hoped against, which is the only realistic stance for any system that absolutely must keep running.

Redundancy comes in forms that differ in how the backup relates to the primary, and the differences matter.

Active redundancy runs multiple units simultaneously, all working at once, so that if one fails the others are already running and carry the load with no interruption; the cost is that you're operating several units all the time.

Standby redundancy keeps a backup unit idle until the primary fails, then switches over to it; this saves the wear and operating cost of running the backup continuously, but the switchover takes time and can itself fail, leaving a gap when the primary quits before the backup is fully online. The choice between them depends on whether you can tolerate any interruption at all, in which case you need active redundancy with its higher cost, or whether a brief switchover gap is acceptable, in which case standby redundancy is cheaper.

A particularly clever form handles not just failures but disagreements. In **voting redundancy**, several units all perform the same computation and a voter compares their outputs, taking the majority answer. With three units, if one produces a wrong answer while the other two agree, the voter takes the agreed answer and the faulty unit is outvoted, so the system tolerates not only a unit dying but a unit lying, producing a plausible but wrong result. This matters enormously in systems where a wrong answer is as dangerous as no answer, like flight control, where a sensor that fails by reporting a confidently incorrect value could be more dangerous than one that simply stops, and the voting scheme catches the liar by consensus.

The principle that makes all redundancy worthwhile, and the trap that makes it fail, is independence. Redundancy only helps if the redundant units fail independently, for unrelated reasons, so that the chance of all of them failing at once is the product of their individual failure chances, which is very small.

The catastrophic trap is **common-cause failure**, a single underlying cause that fails all the redundant units together, defeating the redundancy entirely. Three redundant units in the same room are vulnerable to a fire in that room. Three running on the same power supply die together when the power fails. Three executing the same software all fall to the same bug, present identically in every copy. The redundancy that looked robust because three independent units would all have to fail becomes worthless when one common cause takes out all three at once, and this is the single most important thing to understand about redundancy: adding more identical copies protects against random independent failure but does nothing against a common cause, and may even create a false confidence that's worse than honest vulnerability.

The most sophisticated answer to common-cause failure is to make the redundant units genuinely different, an approach called **dissimilar redundancy** or design diversity. If redundant units share a design, they share its flaws, so a flaw in the design appears in every copy and a common cause defeats them all. Making the units different in design means a flaw in one isn't necessarily present in the others, so a common cause that exploits one design's weakness doesn't take down a different design that lacks that weakness.

The Boeing 777's flight control computers used this principle: the redundant computers used different processors and software developed by different teams, so a bug or a hardware flaw in one design wouldn't be replicated in the others, and a common cause rooted in a design defect couldn't fail all of them together. The cost is enormous, building several genuinely different versions of the same function is far more expensive than copying one version, but for a system where common-cause failure of redundant units would be catastrophic, the diversity is what makes the redundancy real rather than illusory.

There's a sobering truth about redundancy that the discipline insists on, which is that it always costs and never fully eliminates risk. Every redundant unit adds weight, cost, complexity, and its own failure modes, and the switching and voting machinery that manages redundancy can itself fail, sometimes becoming a new single point of failure if it's not designed carefully. A system can be made more reliable through redundancy, but never perfectly reliable, because there's always some common cause, however remote, that could defeat the redundancy, and the cost of each additional layer of protection rises while the risk it removes shrinks. The engineering judgment is in deciding how much redundancy is enough: enough that the remaining risk is acceptable for what the system does, not so much that the cost and complexity become their own source of failure. A passenger aircraft justifies extensive redundancy because lives depend on it; a toaster justifies none. Matching the redundancy to the stakes, and understanding that it reduces rather than removes risk, is what separates sound fault-tolerant design from either reckless fragility or paranoid over-engineering that fails under its own complexity.

Human Factors in System Failure

Every system involves people, and people are part of the system, not separate from it. **Human factors** is the discipline of understanding how people interact with systems and designing those systems to work well with actual human capabilities and limitations rather than with some idealized perfect operator who never tires, never misreads a display, and never makes a mistake. Ignoring human factors is one of the most common and most preventable causes of system failure, because a system that's technically sound can still fail catastrophically if it sets its human operators up to err.

The mistake that has to be undone first is the instinct to blame the operator. When a system fails and a human did something wrong, the easy conclusion is human error, the operator made a mistake, and the fix is to tell people to be more careful or to fire the one who erred. This conclusion is usually wrong, or at least incomplete, because it treats the error as a personal failing rather than as a predictable result of how the system was designed. Most human errors aren't random acts of carelessness; they're the natural consequence of a design that made the error easy to make and hard to catch. A control that's easy to confuse with another, a display that's hard to read under pressure, an alarm that cries wolf so often that operators learn to ignore it, a procedure that demands inhuman vigilance, these are design choices that produce human errors as reliably as a flawed circuit produces electrical faults. Blaming the operator hides the real cause, which is the design, and guarantees the error will happen again with the next operator, because the trap is still there.

Three patterns of human-factors failure show up again and again, and recognizing them is the start of designing around them.

Mode confusion happens when a system can be in different modes that behave differently, and the operator loses track of which mode it's in, so an action that's correct in one mode is dangerous in another and the operator, thinking the system is in the first mode when it's actually in the second, does the wrong thing while believing they're doing right.

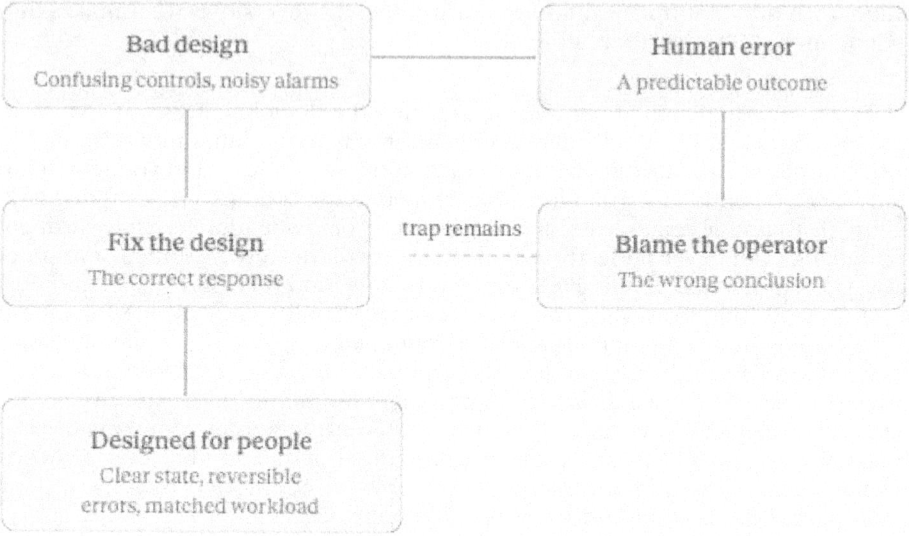

The operator is a component, not an afterthought

Alarm fatigue happens when a system produces so many alarms, many of them false or trivial, that operators become desensitized and start ignoring or silencing them, so that when a real, serious alarm sounds it's lost in the noise of all the ones that didn't matter, and the operators miss the warning that could have prevented disaster precisely because the system had trained them to disregard warnings.

Automation complacency happens when a system is automated enough that operators come to trust it and stop paying close attention, losing the situational awareness and the practiced skill they'd need to take over when the automation fails or encounters something it can't handle, so that at the critical moment when human judgment is most needed, the human is least prepared to provide it.

All three turned correct intentions into wrong outcomes

The accident that turned human factors from a niche concern into a recognized discipline was the partial meltdown at **Three Mile Island** in 1979. The plant suffered a mechanical problem, a valve stuck open, that should have been manageable, but it escalated into a partial core meltdown largely because of how the control room was designed and how the operators interpreted it. A central failure was an indicator that didn't show what operators thought it showed: a light indicated that a relief valve had been commanded closed, and operators read this as confirmation that the valve was closed, when in fact the valve was stuck open and the light only reflected the command sent to it, not the valve's actual position. Believing the valve was closed when it was open, the operators made decisions that were correct for the situation they thought they were in and wrong for the situation they were actually in, and the reactor lost coolant while they acted on a false picture of its state. The control room compounded this with a flood of alarms, hundreds going off, with no clear way to tell which mattered, overwhelming the operators exactly when they most needed clarity. The investigation concluded that the operators had behaved reasonably given the information they were given, and that the real failures were in the design of the human-machine interface, the misleading indicator, the alarm overload, the controls that didn't support the operators in understanding what was happening. The accident reshaped how control rooms are designed, and it established the principle that an interface that misleads its operators is a design defect, not an excuse to blame the people who were misled.

Designing for human factors means building systems that fit how people actually think and behave, and the principles are concrete. Make the system's state clear, so operators always know what mode it's in and what it's actually doing, not just what it was told to do. Make errors hard to commit and easy to reverse, so that a slip doesn't become a catastrophe and a wrong action can be undone before it causes harm. Reduce alarms to the ones that matter and make their priority clear, so that a serious warning stands out instead of drowning in trivia. Match the workload to human capacity, neither overwhelming operators with more than they can process in a crisis nor leaving them so idle and complacent that they lose the alertness they'll need.

And design the system around the operator's actual mental model of how it works, so that the system behaves the way the operator expects, because surprises are where errors live. These principles treat the human as a component with known capabilities and limitations, to be designed for as carefully as any other component, rather than as an infinitely adaptable afterthought expected to compensate for whatever the rest of the design failed to handle. A system designed this way doesn't eliminate human error, which is impossible, but it stops manufacturing errors through bad design, and it catches the errors that do happen before they become disasters.

Margin and Robustness Design

The final piece of risk and reliability thinking is about building systems that survive conditions worse than expected, and it rests on a humble admission: you don't actually know exactly what your system will face, so you'd better build in room to spare.

Margin is the gap between what a system can handle and what it's expected to encounter, the extra capacity beyond the requirement, and **robustness** is the broader quality of performing well across a range of conditions including ones the designers didn't precisely anticipate. A system with margin and robustness survives the surprises that reality always delivers, while a system designed to exactly meet its requirements with nothing to spare fails the moment reality exceeds the assumptions, which it eventually does.

Margin shows up wherever engineers design against a load, and the **safety factor** is its most familiar form: you design a structure to handle some multiple of the maximum load you expect, not just the expected load itself. A cable expected to bear a certain weight might be built to hold several times that weight, so that unexpected overloads, gradual weakening, material flaws, and the simple fact that your estimate of the maximum load might be too low are all absorbed by the extra capacity. The safety factor exists precisely because the designer admits ignorance: the real maximum load is uncertain, the material's real strength varies, the future holds conditions you can't foresee, and the margin is the protection against all of that uncertainty at once. A bridge built to exactly hold its expected traffic with a safety factor of one would be a bridge waiting to fall, because the day will come when the load exceeds the estimate or the steel is slightly weaker than assumed, and with no margin, that day is the day it collapses.

The philosophical heart of robustness is designing for uncertainty rather than for a precise prediction. If you knew exactly what conditions a system would face, you could design it to meet exactly those conditions and nothing more, and it would be optimal.

But you never know exactly, so designing for a precise prediction means designing for conditions that will turn out to be wrong, and a system optimized for the wrong conditions can fail badly when the real conditions differ even slightly. Robustness accepts that the future is uncertain and builds a system that performs acceptably across a range of possible conditions rather than optimally for one guessed condition, trading peak performance under ideal assumptions for resilience under the messy reality that actually arrives. This is why the most robust design is rarely the most optimized one: optimization tunes a system tightly to assumed conditions, and that tightness becomes fragility when the assumptions break, while robustness deliberately leaves the system less perfectly tuned so it can absorb conditions outside the assumptions without failing.

There's a real tension here, because margin and robustness cost something, and the pressure to remove them is constant. Margin is, by definition, capacity you're paying for and not using under normal conditions: the extra steel, the oversized component, the unused performance headroom all add weight and cost, and they look like waste to anyone examining the system under normal operation, where the margin sits idle and apparently unnecessary. So there's perpetual pressure to trim margins, to optimize closer to the expected conditions, to remove the "excess" capacity in the name of efficiency and cost savings. Sometimes this trimming is justified, when a margin really is excessive and the conditions are well enough understood to safely reduce it. But trimming margins also removes exactly the protection that handles the surprises, and a system optimized down to its expected conditions has no reserve left for the day reality exceeds them. The discipline is in knowing which margins are genuinely excess and which are the protection that will save the system, and resisting the pressure to strip out the protection along with the waste, because the margin you remove to save cost is the margin you'll wish you had on the worst day.

The Apollo program, which opened this chapter with a near-disaster, also showed what margin and robustness can do when they're preserved. The systems that brought Apollo 13's crew home were able to be improvised into life-saving configurations partly because they had margin and flexibility designed in, capacity and adaptability beyond their nominal requirements that the crew and ground controllers could press into uses no one had planned. The lunar module, designed to land two men on the Moon, became a lifeboat for three on the way home, running far longer than intended and powering systems it was never meant to power, because it had margin to give. A system designed to exactly meet its original requirements with nothing to spare couldn't have been adapted that way, because there'd have been no reserve to draw on when the situation demanded something the designers never anticipated.

The margin that looked like excess under normal conditions became the difference between three astronauts living and dying when conditions went catastrophically off plan, which is the entire argument for margin and robustness in one example: you build in more than you need precisely because the day will come when you need more than you planned for, and on that day, the margin is everything.

Discussion question: If margin and robustness protect a system against the conditions its designers failed to anticipate, but every margin is also a cost that efficiency pressures will push to eliminate, then how can an organization decide how much protection against the unknown is worth paying for, and who should bear responsibility when a margin that was trimmed for savings turns out to have been the one thing that could have prevented a disaster?

Chapter 6: Integration, Verification, and Validation

In 1994, the European Space Agency was finishing its new Ariane 5 rocket, a heavy launcher that represented a decade of work and billions of euros. Much of its control software was reused from the older, smaller Ariane 4, which made sense: the code was proven, it had flown successfully many times, and rewriting working software invites new bugs. On June 4, 1996, Ariane 5 launched for the first time. Thirty-seven seconds later it veered off course and destroyed itself, along with four expensive satellites, in one of the most costly software failures ever recorded. The cause was a piece of reused code that converted a number from one format to another. On Ariane 4 the value being converted was always small enough to fit. Ariane 5 flew faster, the value grew larger than the old code could hold, the conversion failed, and the failure cascaded through the guidance system until the rocket tore itself apart. The code worked perfectly. It had been verified thoroughly. But it had never been validated against the actual flight conditions of the new rocket, because everyone assumed proven Ariane 4 code would work on Ariane 5, and the integration testing that would have caught the mismatch was skipped for that component.

This chapter is about the stage where separately built pieces come together and you find out whether they actually work, both individually and as a whole, and whether the whole thing solves the problem it was meant to solve. Integration is the act of assembling the pieces. Verification asks whether you built the system right, whether each piece meets its specification. Validation asks the deeper question of whether you built the right system, whether the thing you built actually meets the real need. The Ariane 5 disaster sits at the intersection of all three: pieces integrated without adequate testing, code that was verified against its old spec but never validated against new conditions, a failure that only the assembled system under realistic conditions would have revealed. This is where projects discover whether years of work add up to something that functions, and it's where the cost of every earlier shortcut finally comes due.

Integration Strategies for Complex Subsystems

Integration is the process of combining separately developed components and subsystems into a working whole, and the way you sequence that combining matters enormously, because the order in which you bring pieces together determines how easily you can find and fix the problems that integration always reveals. Bring everything together at once and a failure could be anywhere, hidden in the tangle of newly connected parts. Bring pieces together gradually in a planned sequence and each new failure points to whatever you just added, which is the difference between debugging that converges and debugging that flails.

The most naive approach, and the one that teaches why strategy matters, is **big bang integration**, where you build all the components separately and then connect them all at once, hoping the assembled system works. The appeal is that it seems to save time, no gradual buildup, just assemble everything and go. The reality is usually a disaster, because when the big-bang system fails, and it will, the failure could originate in any component or any interface among them, with no way to isolate the cause. You're left staring at a complex assembly that doesn't work, unable to tell whether the problem is in component A, component B, the connection between them, or some interaction among a dozen parts, and every fix you try might break something else. Big bang integration trades a little apparent time savings up front for enormous debugging pain later, and it's the approach to avoid on anything complex.

The disciplined alternatives bring pieces together gradually, and they come in two main directions.

Top-down integration starts with the high-level structure and adds lower-level components progressively, using temporary stand-ins called stubs to represent the pieces not yet integrated, so the overall framework gets tested early while the details are filled in over time.

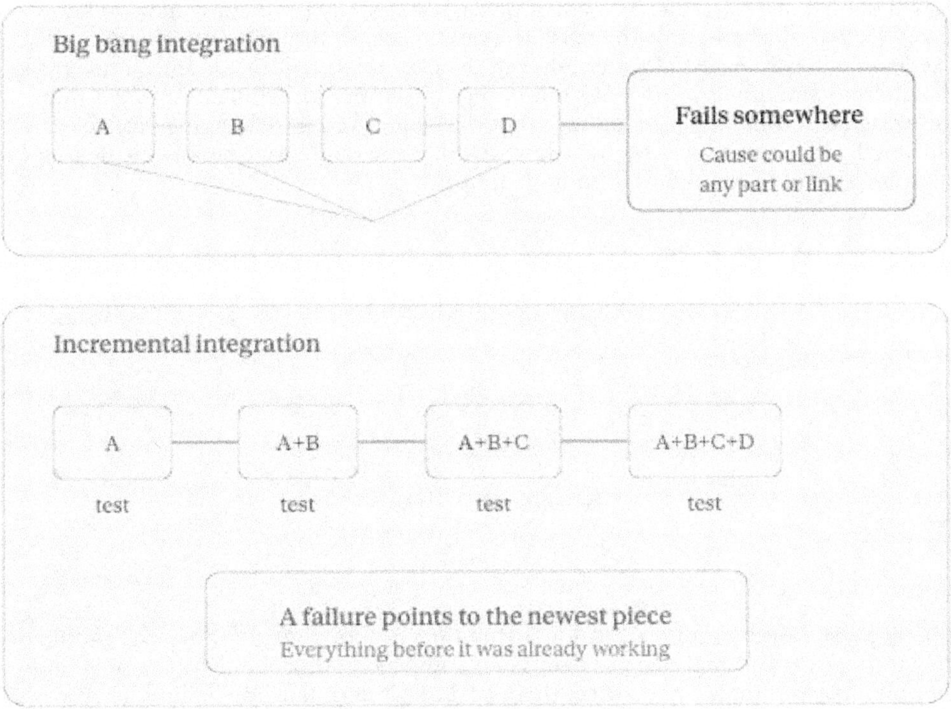

(The top shows big bang integration, where all four components connect at once and a failure could trace to any of them or their links. The bottom shows incremental integration (covered more on the next page) adding one component at a time, testing after each step, so a failure always points to the piece just added.)

Bottom-up integration starts with the lowest-level components, tests them together first, and builds upward toward the complete system, using temporary drivers to exercise the low-level pieces before the higher levels exist. Each has trade-offs: top-down tests the overall architecture and major interfaces early, which surfaces big structural problems sooner, but it needs stubs to stand in for the unbuilt lower pieces and can leave detailed component testing until late. Bottom-up tests the foundational pieces thoroughly first, which builds on solid ground, but it leaves the overall system behavior until the end, so architectural problems surface late. Real projects usually blend the two, working from both ends toward the middle, rather than committing purely to one direction.

The strategy that's become standard for complex systems is **incremental integration**, adding components one or a few at a time and testing after each addition, so that the system grows in small, verified steps. The power of this approach is in fault isolation: when you add one component and something breaks, the problem is almost certainly in the component you just added or in its connection to what was already working, because everything else was working a moment ago before you added the new piece. This narrows the search for the cause from "somewhere in the whole system" to "in or around the thing I just integrated," which makes debugging tractable. You add a piece, test, fix any problem while you know exactly where it is, and only then add the next piece, so problems are caught and fixed while they're easy to locate rather than accumulating into a tangle. The cost is that incremental integration takes more steps and more individual test cycles than throwing everything together at once, but the time saved in debugging vastly outweighs the extra test cycles, which is why incremental integration is the default for anything complex enough that finding faults is hard.

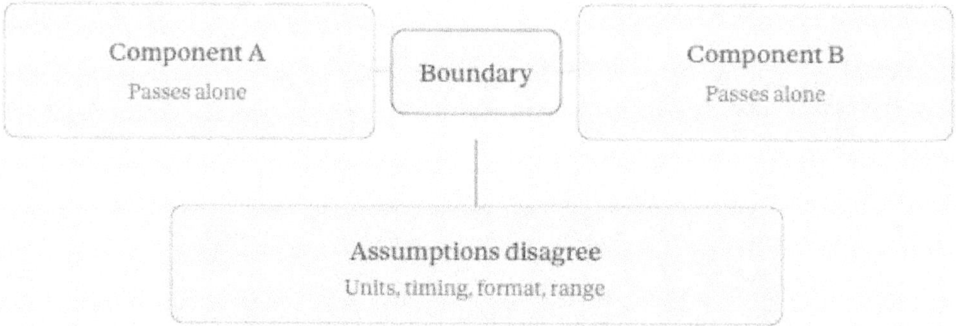

The deepest lesson about integration is that the components are rarely the problem; the interfaces are. When you've tested each component thoroughly on its own and they all pass, and then you connect them and the system fails, the failure usually lives in the connections, the places where the components have to agree about what they're exchanging, not in the components themselves. This is the same truth that runs through systems engineering, surfacing again at the moment of assembly: well-built parts fail to work together because their understandings of the boundary between them don't quite match, in a data format, a timing assumption, a unit, a sequence, a shared resource. Ariane 5 is exactly this, a piece of code that worked perfectly within itself but failed at the boundary where its assumptions about the input range met a reality those assumptions didn't cover. Integration is where these interface mismatches finally become visible, because it's the first time the components actually have to interact, and a great deal of integration's purpose is precisely to flush out the interface problems that component testing can't catch, since component testing checks each piece against its own spec while integration checks whether the pieces' specs actually fit together.

Test Planning and Test Matrices

You can't test everything, and understanding why not is the start of testing well. The number of possible conditions a real system can encounter, every combination of inputs, states, environments, sequences, and timings, is so vast that testing all of them would take longer than the life of the universe for even a modest system. So testing is always a process of choosing which of the effectively infinite possible tests to actually run, and the quality of that choosing determines whether your finite testing actually finds the problems that matter or wastes effort on conditions that don't while missing the ones that do.

A **test plan** is the document that defines what will be tested, how, and against what criteria, and its real value is that it forces these choices to be made deliberately and in advance, rather than testing haphazardly and hoping you covered the important cases. The plan specifies what gets tested, what conditions each test runs under, what result counts as passing, and who's responsible for each test. Writing it before testing begins means the choices about what to cover are made thoughtfully, with the whole picture in view, instead of being made implicitly by whatever the testers happened to think of in the moment, which inevitably leaves gaps where nobody thought to look.

A **test matrix** organizes this coverage into a structured grid that makes the gaps visible, typically mapping test cases against the requirements or conditions they cover, so you can see at a glance what's tested and, more importantly, what isn't. The matrix's main job is to expose holes in coverage, the requirement that no test exercises, the condition that nothing checks, the combination that falls between the tests you planned. Laid out as a grid, these gaps jump out, where a requirement with no test in its row is a promise nobody's verifying and a condition with an empty column is a risk nobody's checking. Without the matrix, coverage gaps hide, because it's easy to believe you've tested thoroughly when you've actually tested only the cases that came naturally to mind and left whole categories untouched. The matrix turns the vague feeling of "we tested a lot" into the concrete question of "what specifically is and isn't covered," which is the only honest basis for deciding whether testing is sufficient.

Because you can't test everything, the art is in choosing tests that find the most problems for the effort, and a few principles guide that choice. Test the **boundaries**, the edges of the valid ranges, because failures cluster at boundaries far more than in the comfortable middle of a range: the input at exactly the maximum, one above it, one below it, the moment a buffer fills, the transition from one state to another. The Ariane 5 failure was a boundary failure, a value that exceeded the range the code could hold, exactly the kind of edge case that boundary testing targets. Test the conditions most likely to occur, so the system is proven where it'll actually spend its time. Test the conditions whose failure would be most severe, so the dangerous cases are checked even if they're rare. And test the unusual combinations and sequences that components might not handle well together, since those interface interactions are where integration failures hide. These principles concentrate finite testing effort where problems actually live, which is far more effective than testing a uniform spread of conditions or, worse, testing mostly the easy normal cases that are least likely to fail.

There's a hard truth about testing that has to be stated plainly, because misunderstanding it causes real harm: testing can show that a system has faults, but it can never prove the system has none. The computer scientist Edsger Dijkstra put it exactly, that testing can show the presence of bugs but never their absence. When testing finds a fault, you've learned something definite, there's a problem, here it is.

When testing finds no fault, you've learned only that the specific conditions you tested didn't reveal a problem, which says nothing about the conditions you didn't test, where faults may well lurk. Passing all your tests means your tests didn't find anything, not that there's nothing to find, and the conditions you didn't think to test are precisely where the surprises wait. This is why "it passed all the tests" is never the same as "it works," and why thoughtful test planning matters so much: since you can't test everything and passing tests doesn't prove correctness, the value of testing depends entirely on whether you tested the right things, the conditions where problems actually live, rather than a comfortable set of cases that pass easily and prove little.

Verification Methods: Test, Analysis, Inspection, Demonstration

Testing is the method everyone thinks of, but it isn't the only way to verify that a system meets its requirements, and using only testing is both wasteful and sometimes impossible.

Verification is the work of confirming that a system meets its specified requirements, that you built the thing right, and there are four recognized methods for doing it, each suited to different situations. Knowing all four, and when to use each, is what lets you verify everything that needs verifying without wasting effort testing what could be confirmed more cheaply and without being stuck when testing simply can't be done.

The four methods are test, analysis, inspection, and demonstration, and the distinctions are practical. **Test** means operating the system under controlled conditions and measuring the results against the requirement, the most direct and convincing method, producing hard evidence that the system actually does what's required under the conditions you applied. **Analysis** means using calculation, modeling, or reasoning to verify a requirement without physically operating the system, which you turn to when testing is impossible, impractical, or prohibitively expensive, proving by math and physics what you can't or won't prove by physical test. **Inspection** means examining the system directly, by looking, measuring, or reviewing, to confirm a requirement, which suits things you can simply observe: a dimension, a material, a label, the presence of a feature, checked by direct examination rather than by operating anything. **Demonstration** means showing that the system performs a function under realistic conditions, less rigorous than a measured test but enough to confirm that a capability works, often used for functions where seeing it happen is sufficient and precise measurement isn't the point.

Choosing among them is a matter of matching the method to the requirement, and the choice often comes down to cost and feasibility. A requirement about a physical dimension is verified by inspection, you measure it, because operating anything would be pointless when looking suffices. A requirement about how a structure responds to a load far beyond what you'd dare apply to the real article is verified by analysis, because testing it to destruction would destroy the thing you need, so you calculate instead. A requirement about response time under load is verified by test, because you can actually run the system and measure how fast it responds, producing direct evidence. A requirement that a user can complete a task is verified by demonstration, you show someone doing it, because the point is that the capability works, not that it's measured to three decimals.

Using the wrong method wastes effort or fails entirely: testing something that should be inspected wastes resources operating a system when looking would do, while trying to test something that can't be safely tested leaves you stuck when analysis would have proven it cleanly.

The most consequential use of analysis over test is for the conditions you can't afford to actually create, and spacecraft show this starkly because their real operating environment is impossible to fully reproduce on the ground. You cannot put a complete spacecraft in true deep-space conditions before launch, the real vacuum, the real radiation, the real thermal extremes, the real zero gravity over months, all at once, because no ground facility reproduces all of it simultaneously and the only place those conditions truly exist is in space itself, where you can't go to test before you've launched.

So spacecraft verification leans heavily on analysis: engineers calculate and model how the spacecraft will behave in conditions they can't physically create, verifying by physics and reasoning what they can't verify by test, and confirming through analysis that the vehicle will survive an environment they're unable to fully reproduce before committing it to that environment for real. Where they can test, they do, in vacuum chambers, on vibration tables, under radiation sources, each reproducing part of the environment, but the complete combined reality of spaceflight is verified through analysis because the alternative, testing in actual deep space, is exactly what launching the spacecraft is, with no second chance if the analysis was wrong. This is why analysis is indispensable rather than a poor substitute for testing: for the conditions that matter most and can't be created on the ground, analysis is the only verification available, and the quality of that analysis is what stands between a working spacecraft and an expensive failure in a place no one can reach.

Validation Against Operational Need

Verification and validation sound similar and get confused constantly, but the difference between them is one of the most important distinctions in all of engineering, because a system can pass every verification and still be a complete failure. **Verification** asks whether you built the system right, whether it meets its specifications. **Validation** asks whether you built the right system, whether it actually meets the real operational need, whether it solves the problem the people who needed it actually have. The two questions are genuinely different, and a system can answer one perfectly while failing the other entirely.

The way to feel the difference is to picture a system that's perfectly verified and useless. Suppose the requirements themselves were wrong, capturing the wrong need, specifying a system that meets every written requirement but doesn't actually solve the user's problem. You build it, you verify it exhaustively, every requirement is met, every test passes, the system does precisely what the specification said. And it's worthless, because the specification described the wrong thing, and a faithful implementation of the wrong specification is a faithful implementation of the wrong system. Verification confirms you matched the spec; it says nothing about whether the spec was right.

This is why validation is a separate and essential question: all the verification in the world only proves you built what you wrote down, and if what you wrote down was wrong, you've built the wrong thing correctly, which is no comfort to the people whose actual need went unmet. Validation is the check against reality that verification can't provide, because validation goes past the specification to ask whether the system serves the genuine purpose, while verification stops at the specification's edge.

When validation gets skipped, the result is a system that works as specified and fails in use, and the history of large projects is full of them: systems delivered that met their requirements and were rejected by the people they were built for because they didn't fit how the work was actually done, didn't solve the real problem, or solved a problem that had been misunderstood from the start. These systems often pass verification with flying colors, because they were built carefully to their specifications, and then fail validation completely, because the specifications never captured the real need. The failure traces all the way back to requirements that were wrong, and verification couldn't catch it because verification only checks against those same wrong requirements, faithfully confirming a system that was doomed from the moment the need was misunderstood.

The London Ambulance dispatch system met its specifications and couldn't handle real dispatching. Healthcare.gov's pieces met their specifications and couldn't serve real users at real scale. The pattern repeats because organizations treat verification as if it were validation, assuming that a system which passes its tests must therefore be a system that works, when passing tests only proves conformance to a spec that may itself be the problem.

Because validation checks against the real need rather than the written spec, it has to involve the real users in realistic conditions, which is what distinguishes it from verification's controlled measurements. You validate by putting the system, or a version of it, in front of the people who'll actually use it, doing the work they actually do, in conditions resembling reality, and seeing whether it genuinely serves them, because the only authority on whether a system meets the real need is the reality of the need itself, embodied in the users and their work. This is why approaches that get real systems in front of real users early, iterative methods that deliver working pieces and gather genuine feedback, are so effective at validation: they test against actual need continuously rather than waiting until the end to discover that a fully built, fully verified system doesn't fit. A system validated only at the very end, after everything's built to spec, risks discovering too late that the spec was wrong, when fixing it means rebuilding rather than adjusting, while a system validated throughout, checked against real users and real work as it develops, catches the gap between spec and need while it's still cheap to close.

The hardest truth in this whole area is that the requirements themselves can be wrong, and validation is the only thing that catches it. Verification trusts the requirements completely, it checks the system against them and asks no questions about whether they're right, so if the requirements are flawed, verification will confirm a flawed system without a hint of trouble. Validation is the safeguard against wrong requirements, the check that goes around the specification to ask whether the whole thing actually serves its purpose, and it's the only safeguard, because nothing else in the process questions whether the requirements captured the real need. A team that verifies thoroughly but validates poorly can build a flawless implementation of a mistake and have every document in order, every test passed, every requirement met, and a system the users can't use.

The Ariane 5 code was verified against its requirements and met them; what failed was the validation against the actual operational reality of a different rocket, the check against the real conditions the system would face, which no amount of verification against the old requirements could substitute for. This is why the discipline insists on both, and why confusing them is so dangerous: verification keeps you honest about meeting your specifications, but only validation keeps you honest about whether your specifications, and the system built from them, actually solve the real problem for the real people who have it.

Discussion question: If verification can only check a system against its requirements, and validation against real users is the only way to discover that the requirements themselves were wrong, then for a system where realistic validation is impossible before commitment, like a spacecraft, a weapon, or any system whose true operating conditions can't be reproduced beforehand, how can a team ever gain real confidence that they built the right thing and not just a flawless version of a misunderstanding?

Chapter 7: Program Management and Systems Engineering

The Sydney Opera House, the F-35 fighter, the Boston Big Dig, the Berlin Brandenburg Airport: the list of famous projects that came in years late and billions over budget is long, and what's striking is how rarely the cause was a technical impossibility. The engineering usually worked, eventually. What broke was the management of the engineering, the structures that were supposed to keep hundreds or thousands of people, working over years, building toward the same target on a budget and a schedule that meant something.

Berlin's new airport opened in 2020, nine years late, after a planned 2011 opening, with costs that ballooned from around two billion euros to over six billion, and the failures that delayed it were things like fire safety systems that didn't work, cabling that wasn't documented, and a sprawling project nobody could fully see or control. The technology to build an airport has existed for a century. What failed in Berlin was the management.

This chapter is about the part of systems engineering that isn't about physics or design but about controlling the work itself, the structures and disciplines that let a large technical effort stay coherent, visible, and on track over the long haul. It covers how you break a huge project into manageable pieces, how you keep track of what version of everything you're actually building, how you measure technical progress before it's too late to course-correct, how you make the hard trade-offs among cost, schedule, and performance, how you use formal checkpoints to catch problems while they're still fixable, and how you manage the accumulated shortcuts and compromises that quietly weigh a project down. None of this is glamorous, and all of it is what separates the projects that deliver from the ones that become cautionary tales.

A recurring lesson in program management is that bad news does not become less expensive by arriving late. Projects almost never fail without warning; the warning signs are usually visible months or years beforehand in slipping schedules, rising defect counts, missed milestones, unstable requirements, or performance measures that quietly stop improving. The danger is that large programs can generate so much information that real problems become buried beneath status reports that look reassuring on the surface. Effective program management is therefore not just about collecting data but about creating visibility, making sure decision-makers see the true state of the project early enough to act. A project that discovers its problems while they are still small often recovers. A project that discovers them only after they have accumulated for years is usually left choosing between delay, massive cost growth, or outright failure.

Work Breakdown Structures

A large project is too big to manage as one thing. You can't plan, schedule, budget, or track "build the aircraft" as a single undertaking, because it's an undertaking made of thousands of smaller pieces of work, and you need a way to divide it into parts small enough to actually plan and assign. The **work breakdown structure**, or WBS, is that division: a hierarchical decomposition of all the work in a project into progressively smaller, more manageable pieces, breaking the whole down into major parts, then breaking those into subparts, and so on until you reach pieces small enough to estimate, assign, and track individually.

The structure is a tree. At the top sits the entire project. Below it are the major components or phases. Below those are smaller pieces, and below those smaller still, each level breaking its parent into the parts that make it up, until you reach the bottom level of pieces small enough to handle as units.

These smallest pieces are called **work packages**, and they're the point of the whole exercise: a work package is a chunk of work small enough that you can put a single owner on it, estimate how long it'll take and what it'll cost, schedule it, and track whether it's done. Above the work packages, the higher levels organize and group, but the work packages are where planning and tracking actually happen, because they're the units small enough to be real.

The discipline that makes a WBS sound is a rule that sounds pedantic and matters enormously: the breakdown must be complete and non-overlapping, often summarized as mutually exclusive and collectively exhaustive. Collectively exhaustive means the pieces, taken together, cover all the work, with nothing left out, because any work that doesn't appear somewhere in the WBS is work that nobody planned, nobody scheduled, nobody budgeted, and nobody's tracking, which means it'll surprise the project later as unplanned effort that blows the schedule and budget nobody allocated for it. Mutually exclusive means the pieces don't overlap, each piece of work belongs to exactly one place in the structure, because work that appears in two places gets either done twice, wasting effort, or assumed to be the other group's job by both groups, so it falls through the gap and doesn't get done at all. A WBS that violates either rule, that leaves gaps or creates overlaps, is a WBS that will produce exactly the surprises and dropped work it was supposed to prevent.

The most common and costly mistake in building a WBS is forgetting the work that isn't a physical deliverable, and this trips up even experienced teams because the invisible work doesn't announce itself. A WBS naturally captures the tangible things, the components you can point to, the parts that get built, but a huge fraction of a real project's effort goes into work that produces no physical object: integration, the labor of making the pieces work together; testing, the effort of verifying everything; documentation, the writing that has to happen; project management itself, the ongoing work of running the project; training, the effort of teaching people to use and maintain the system. This work is real, it consumes enormous time and money, and it's easy to leave out of a WBS because it's not a thing you can hold, so the WBS captures all the deliverable components and quietly omits the integration and testing and management that may cost as much as the components themselves. A project planned from a WBS that omits this invisible work is a project that's underestimated its effort badly, because the missing work doesn't go away, it just isn't in the plan, and it surfaces as overruns when the project hits the integration and testing it never budgeted for.

The WBS earns its place by becoming the backbone that everything else hangs on, which is why it's built first. The schedule is built from the work packages, each one timed and sequenced. The budget is built from the work packages, each one costed and summed up the tree. Responsibility is assigned by work package, each one owned by someone. Progress is tracked by work package, each one marked done or not. The WBS is the common structure that ties the schedule, the budget, the assignments, and the tracking together, so that they all refer to the same pieces of work and stay consistent with each other, rather than being separate views that drift apart. A change to the work shows up in the WBS and flows out to the schedule, the budget, and the tracking, all referring to the same decomposition. Get the WBS right, complete and non-overlapping, capturing all the work including the invisible kind, and you have a sound foundation for managing the whole project. Get it wrong, with gaps and overlaps and missing work, and every plan built on it inherits those flaws, which is why the WBS comes first and why the discipline of building it carefully repays itself many times over across the life of the project.

Configuration Management

On a large project, thousands of items, documents, designs, software versions, parts, are changing constantly, and a question that sounds trivial becomes genuinely hard: what, exactly, are we building right now? Which version of each design, each document, each piece of software, each part is the current, official one that everyone should be working from?

Configuration management, or CM, is the discipline of answering that question, of knowing and controlling exactly what version of everything makes up the system at any given time, so that the thousands of changing items stay coordinated rather than dissolving into confusion about which version is real.

The problem CM solves is vicious precisely because it's mundane. When many people change many items over a long time, it becomes shockingly easy to lose track of which version is current, and the failures that result are maddening: someone builds against an old version of a design that's since changed, someone fixes a bug in one copy of the software while another copy still has it, two teams work from two different versions of an interface document and build incompatible parts, a part gets manufactured to a drawing that was superseded months ago. Each of these is a version-control failure, work done against the wrong version of something because nobody had firm control over which version was official, and on a large project these failures multiply until people genuinely don't know what they're building, with different groups holding different versions of the truth and discovering the mismatches only when the incompatible pieces meet.

CM rests on a small set of concepts that together impose order. A **configuration item** is anything under configuration management's control, any document, design, software component, or part whose version is tracked and controlled. A **baseline** is an officially approved version of an item or set of items at a point in time, a fixed reference that says this is the official version right now, everyone works from this, and changes are measured against it. **Version control** is the tracking of changes over time, keeping the history of each item so you know what changed, when, and why, and can always identify the current version and recover earlier ones. And **change control**, met earlier in the context of requirements, is the process by which changes to baselined items happen deliberately and with approval, rather than casually, so that the official version changes only when someone with authority decides it should. Together these mean that at any moment there's a known, official version of everything, changes to that official version are controlled and recorded, and the history is preserved, which is exactly the order that prevents the version-control failures.

The four functions CM performs are worth naming because they map onto the questions CM exists to answer. Identification means defining what's under control and labeling each item and version clearly, so everything has a known identity. Control means managing changes through a deliberate process, so versions change only by approval. Status accounting means tracking and reporting the current state, what version everything's at right now and what changes are in progress, so the project can always see where things stand. Audit means verifying that the system actually matches its documentation, that what you built is what the official baseline says you built, catching the gaps where reality has drifted from the records.

These four functions, identifying what's controlled, controlling its changes, accounting for its status, and auditing reality against records, are how CM keeps a sprawling, changing project coordinated, and skipping any of them opens the door to the confusion CM exists to prevent.

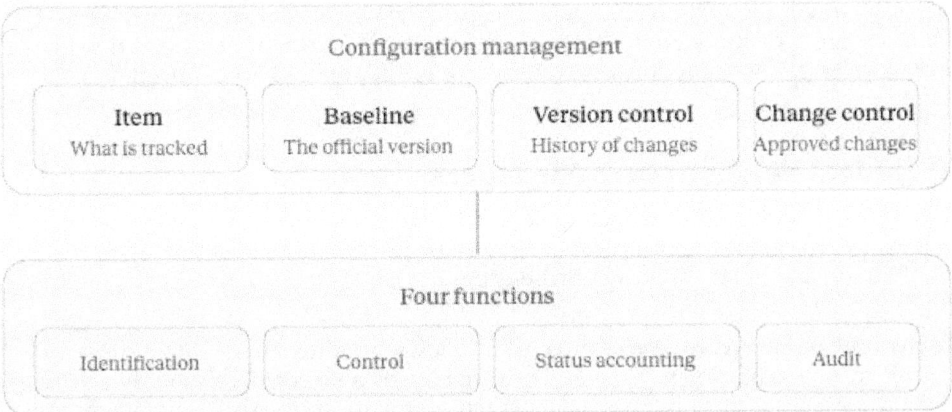

The cost of weak configuration management shows up dramatically in software, where versions multiply fast and the failures are immediate and visible. Modern software development lives or dies by configuration management, because a serious software system has countless versions, components, and dependencies changing continuously, and without firm control over which version of everything goes together, builds fail, bugs reappear after being fixed, and deployments break because the pieces in production don't match the pieces that were tested. The whole modern practice of disciplined version control, where every change is tracked, every version is identified, and the exact configuration of any build can be reconstructed, exists because software taught the industry, painfully and repeatedly, that losing track of versions is catastrophic. A bug gets fixed, then reappears because the fix was made to a version that got overwritten by an older one. A system works in testing and breaks in production because production is running a different combination of component versions than testing was. These failures, common before disciplined CM and largely preventable with it, are why software development built configuration management into the core of how it works, and why the tools for tracking versions and controlling changes are among the most-used tools in the entire field.

The deeper point is that configuration management is invisible when it works and catastrophic when it doesn't, which is why it's chronically underappreciated until a failure makes its absence felt.

When CM is done well, nobody notices, because everything just stays coordinated, the right versions go together, changes happen cleanly, and the system matches its documentation, so there's no drama and no credit. When CM is done poorly, the failures are baffling and expensive, work done against wrong versions, bugs that won't stay fixed, parts that don't match their drawings, incompatibilities discovered at the worst moment, and tracing these failures back to their root cause in poor version control is itself hard, because the symptom, an incompatible part, a recurring bug, looks like a technical problem rather than the configuration-management failure it actually is.

This is why CM is one of the disciplines that experienced engineers respect and beginners overlook: it produces no visible output of its own, it feels like bureaucratic overhead, and its entire value is in the failures that don't happen, which are invisible precisely because they didn't happen. The projects that take it seriously avoid a whole category of maddening, expensive confusion, and the projects that don't pay for the lesson the way the software industry once did, one version-control disaster at a time.

Technical Performance Measures

A project can look healthy on schedule and budget while the system it's building is quietly failing to meet its technical goals, and by the time that becomes undeniable, it's often too late to fix without enormous cost.

Technical performance measures, or TPMs, are the defense against this: specific technical parameters tracked over the life of a project to give early warning when the system is drifting away from its requirements, so the drift gets caught and corrected while correction is still possible rather than after it's become a crisis.

The problem TPMs solve is that schedule and cost tracking, the usual measures of project health, don't tell you whether the system will actually work. You can be on schedule and on budget and heading straight for a system that misses its weight target, or falls short on range, or can't hit its performance requirement, because schedule and cost measure how the work is progressing, not whether the technical result will meet the need. A project that tracks only schedule and budget can sail along looking fine right up until integration or testing reveals that the system doesn't meet its technical requirements, at which point the project that looked healthy is suddenly in deep trouble, facing a technical shortfall discovered too late. TPMs fill this gap by tracking the technical parameters themselves, the actual measures of whether the system will perform, so that technical problems surface early in the tracking rather than late in the testing.

A TPM works by comparing the current value of a parameter against where it should be at this point in the project, and the structure is what gives the early warning. For each tracked parameter you set a final required value, the target the system must hit. You plan a profile of where the parameter should be at each point along the way, recognizing that some parameters start off bad and improve as the design matures, so the planned profile shows the expected progression toward the target. Then you measure the actual value periodically and plot it against the planned profile, watching whether the parameter is tracking toward its target or drifting away from it. The comparison between actual and planned is the warning signal: a parameter tracking close to its planned profile is healthy, while a parameter falling behind its profile, not improving as fast as planned or actively getting worse, is a warning that the system is drifting toward missing its requirement, visible now while there's still time to act rather than at the end when there isn't.

Weight is the classic TPM in aerospace, and it shows exactly why early tracking matters, because weight has a way of growing relentlessly throughout a project as a thousand small decisions each add a little. Every aircraft and spacecraft program tracks weight obsessively, because weight tends to creep upward as the design matures, each component coming in a little heavier than estimated, each change adding a bit more, and the accumulation, if unwatched, pushes the vehicle past its weight limit, which is often a hard limit tied to whether it can fly or reach orbit at all. So programs set a weight target, plan how weight should grow and then be controlled as the design firms up, and track actual weight against that plan continuously, watching for the creep that signals trouble. When the tracked weight starts drifting above its planned profile, climbing toward the limit faster than allowed, the program sees it early and acts, finding weight to remove, redesigning heavy components, attacking the growth while there's still design freedom to do something about it. A program that didn't track weight this way would discover its weight problem at the end, when the vehicle is built and overweight and fixing it means tearing into finished hardware or accepting a vehicle that can't do its job, which is exactly the late, expensive crisis that early TPM tracking exists to prevent.

The real value of TPMs is that they convert a late catastrophe into an early, manageable problem, which is the whole game in managing technical risk. A technical shortfall discovered at the end of a project, in integration or testing, is a catastrophe, because the design is frozen, the hardware is built, and fixing the shortfall means expensive rework or worse. The same shortfall, seen early through TPM tracking as a parameter drifting off its target, is a manageable problem, because there's still design freedom, still time to adjust, still cheap options available before everything's committed. TPMs don't fix technical problems, they reveal them early, and that early revelation is what makes the problems fixable, because the cost of fixing a technical problem rises steeply as the project progresses, the same exponential cost curve that runs through all of systems engineering. By surfacing technical drift while it's still cheap to correct, TPMs let a project steer toward its technical goals continuously rather than discovering at the end that it missed them, which is the difference between a project that adjusts and delivers and one that hits an immovable technical wall after it's too late to do anything but pay.

Schedule and Cost Trade Studies

Every project lives under three pressures that pull against each other: it has to perform well enough, cost little enough, and finish soon enough, and these three almost never align, so improving one usually means sacrificing another. A **trade study** is the structured analysis of these competing options, a disciplined way of comparing alternatives that differ in cost, schedule, and performance so that the choice among them is made deliberately and defensibly rather than by gut feel or by whoever argues loudest. The trade study is how systems engineering makes the hard choices that have no clean answer, by laying out the alternatives and their consequences clearly enough that a reasoned decision becomes possible.

The framework everyone uses to think about these tensions is the **iron triangle**, sometimes called the project management triangle, which captures the relationship among cost, schedule, and performance, often phrased informally as fast, good, and cheap, with the hard truth that you can't maximize all three at once. Want it faster? It'll usually cost more, as you add people and resources to speed things up, or it'll perform worse, as you cut scope to hit the date. Want it cheaper? It'll usually take longer, as you do with less, or perform worse, as you settle for less capability. Want it to perform better?

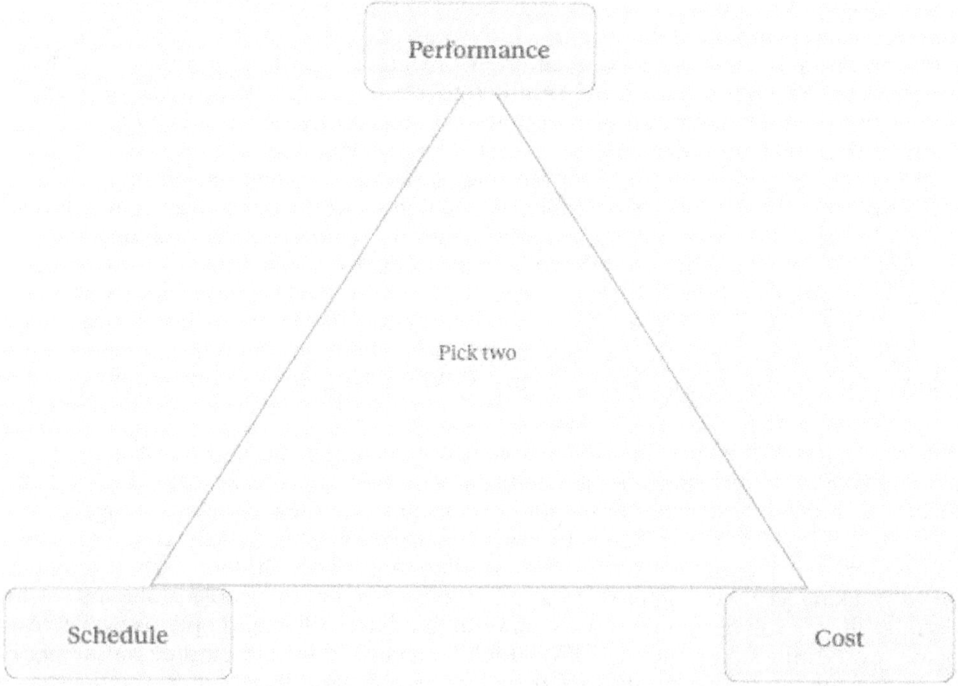

It'll usually cost more and take longer, as you add the effort that higher performance demands. The three corners pull against each other, and the iron triangle names this so that everyone understands the choice isn't whether to sacrifice something but which thing to sacrifice, because some sacrifice among cost, schedule, and performance is nearly always required, and pretending otherwise just means the sacrifice happens by accident rather than by choice.

A trade study makes these choices rigorous through a structured comparison, and the discipline is in doing it explicitly rather than in someone's head. You define the alternatives you're choosing among, the genuinely different options on the table. You establish the criteria that matter, the dimensions on which the alternatives differ and that the decision should weigh, cost, schedule, performance, risk, and whatever else is relevant. You weight the criteria by how much each matters, because not all criteria are equally important and the weighting should reflect the project's actual priorities. You evaluate each alternative against each criterion, scoring how well each option does on each dimension. And you combine the scores and weights to compare the alternatives, producing a structured basis for the decision that shows not just which option comes out ahead but why, in terms of how it performs on the criteria that matter. The output isn't a mechanical answer that makes the decision for you, but a clear, defensible analysis that supports a reasoned choice and, just as importantly, records the reasoning so that the decision can be explained and revisited later.

The danger in trade studies is that their apparent rigor can disguise predetermined conclusions, and this is a genuinely serious trap because it corrupts the whole point of doing the analysis. A trade study produces numbers, scores, weights, totals, and numbers look objective, which makes it tempting and easy to work backward: decide which alternative you want first, then choose the criteria and weights that make that alternative win, and present the result as if the analysis drove the decision when really the decision drove the analysis. This rigged trade study is worse than no trade study, because it wraps a predetermined choice in a veneer of objectivity, giving a decision that was made on bias or politics the false authority of a rigorous analysis, and it's hard to challenge precisely because it looks so methodical.

The criteria and weights are where the manipulation hides, since they're matters of judgment that can be tuned to produce any desired outcome, and a trade study where the weights were chosen to get a particular answer is a justification dressed as an analysis. The defense is honesty about the criteria and weights, setting them based on what genuinely matters before knowing which alternative they'll favor, and being transparent about the judgment involved, so that the trade study informs the decision rather than laundering one that was already made.

Trade studies done with integrity are genuinely valuable, and their value is as much in the record as in the result. A sound trade study forces the alternatives and their consequences into the open, makes the basis for the choice explicit, and documents the reasoning, which matters enormously when decisions get questioned later, as they always do on long projects when circumstances change or things go wrong and someone asks why a particular choice was made. A project with honest trade studies can answer that question, pointing to the analysis that showed the alternatives, the criteria, and the reasoning behind the choice, while a project that made its choices by gut feel has nothing to show but assertions. The discipline of the trade study, comparing alternatives explicitly against weighted criteria and recording the reasoning, is what lets a project make defensible decisions among genuinely hard trade-offs and stand behind those decisions afterward, which is exactly what the cost-schedule-performance tensions demand, because there's no avoiding the trade-offs and the only real choice is whether to make them thoughtfully and on the record or carelessly and in the dark.

Decision Gates and Reviews

A long project needs checkpoints, moments where everyone stops, assesses whether the project is actually ready to continue, and makes a deliberate decision about whether to proceed, because a project that just rolls forward without these checkpoints can carry serious problems all the way to the end before anyone forces a reckoning.

Decision gates, also called milestone reviews or phase gates, are these formal checkpoints: defined points in the project where progress is reviewed and a deliberate go-or-no-go decision is made about whether to move into the next phase, so that the project advances only when it's genuinely ready and problems get confronted at the gate rather than carried past it.

The logic of a gate is that the project can't proceed past it until specific criteria are met, which forces problems to be confronted rather than carried forward, and this forcing function is the entire point. At each gate, the project has to demonstrate that it's actually ready for the next phase, meeting defined entry criteria for that phase, and if it can't, the gate is where that gets exposed and dealt with, rather than the project sliding into the next phase carrying unresolved problems that will surface later, bigger and more expensive. The gate is a moment of honest assessment, a forced pause to ask whether the project is really ready, and its value is precisely in being a barrier that an unready project can't simply roll through, because the alternative, a project that flows continuously from phase to phase without these checkpoints, is a project where problems accumulate silently and surface only at the end, when they're catastrophic.

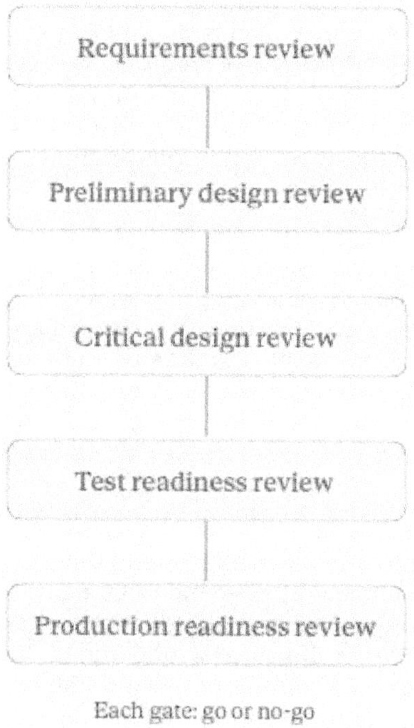

Major projects have a recognizable sequence of gates that track the project's maturity, each one a checkpoint at a meaningful transition. A requirements review confirms that the requirements are complete and sound before the project commits to designing against them. A preliminary design review confirms that the overall design approach is sound before the project commits to detailed design. A critical design review confirms that the detailed design is complete and ready before the project commits to building. A test readiness review confirms that the system and the test setup are actually ready before testing begins. A production readiness review confirms that everything's ready before committing to full manufacturing. Each gate sits at a point where the project is about to commit significant resources to the next phase, and each asks whether the work so far genuinely justifies that commitment, so the gates collectively ensure that the project advances through its phases only as it earns the right to, with each major commitment preceded by a check that the foundation for it is actually in place.

The way a gate review can fail is the most important thing to understand about gates, because a gate that fails this way is worse than no gate at all. A gate review becomes worthless if it turns into a rubber stamp, a checkpoint where the decision to proceed is a foregone conclusion and the review is theater performed to bless a continuation that was never really in question. This happens under schedule pressure, when stopping or slowing the project feels unacceptable, so the gate that's supposed to be an honest go-or-no-go becomes a ritual where everyone knows the answer is go regardless of whether the criteria are actually met, and the review's hard questions get softened or skipped because the project is going to proceed anyway. A rubber-stamp gate is worse than no gate, because it consumes the time and effort of a review while providing none of the protection, and worse, it provides false assurance, letting everyone believe the project passed a real checkpoint when it passed only a ceremony, so the problems the gate should have caught roll forward under the cover of a review that pretended to look for them. The value of a gate is entirely in its willingness to actually stop a project that isn't ready, and a gate that won't stop anything isn't a gate, it's a costume.

This connects to a hard reality about gates that organizations struggle with, which is that a real gate has to be willing to deliver bad news and act on it, including the hardest decision of all, to stop a project that shouldn't continue. The honest assessment a gate demands sometimes reaches an uncomfortable conclusion, that the project isn't ready, or is in worse shape than hoped, or in the extreme shouldn't proceed at all, and a gate that's genuine has to be able to act on that conclusion rather than waving the project through to avoid the discomfort.

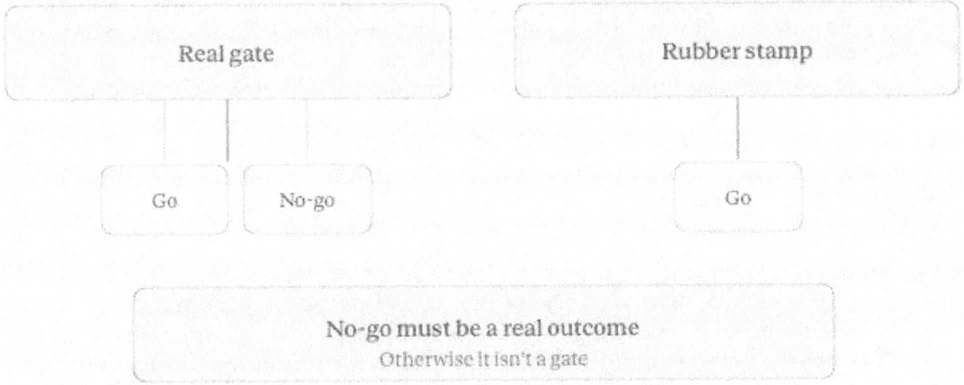

The willingness to stop a project at a gate, to say no-go and mean it, is what makes the entire system of gates worth anything, because the protection gates provide is precisely the protection of not carrying an unready or doomed project forward, and that protection only exists if no-go is a real possible outcome. Organizations that can't bring themselves to stop projects, that treat every gate as a formality on the way to inevitable continuation, lose the whole benefit, because they've kept the form of the checkpoint while removing its substance, and a project that can never be stopped at a gate is a project that will carry its problems all the way to the expensive end, which is the exact outcome the gates were built to prevent.

Managing Technical Debt

When a team takes a shortcut to move faster now, choosing the quick solution over the right one to hit a deadline, they create something that has to be paid back later, and the metaphor that's spread across engineering captures it precisely.

Technical debt is the accumulated cost of choosing expedient shortcuts over sound solutions, the future work created by present compromises, and like financial debt it carries interest, because the shortcuts make everything built on top of them harder until the debt is paid down. The concept matters because technical debt is often invisible, accumulating quietly while the project appears to move fast, until the accumulated weight of it slows everything to a crawl.

The metaphor is worth taking seriously because its parts map so well onto reality. You take on technical debt to gain speed now, just as you borrow money to have it now, choosing a quick compromise that lets you hit a deadline instead of the slower sound solution. The debt accrues interest, meaning the shortcut makes future work harder, because everything built on top of a compromise has to work around it, and the more you build on the shaky foundation the more the compromise costs you in extra effort, work-arounds, and fragility. And the debt has to be repaid eventually through the work of fixing the shortcut properly, or the interest keeps mounting, the compromise keeps making everything harder, until the accumulated burden becomes crippling. Like financial debt, a moderate amount taken on deliberately and paid down responsibly is manageable and sometimes wise, while debt that accumulates unwatched and unpaid grows until it overwhelms the ability to service it.

The honest and slightly surprising truth is that technical debt isn't always bad, and treating it as purely a failure misses how it's actually used well. Sometimes taking a shortcut to hit a deadline is the right call, a deliberate, conscious decision to trade some future cost for speed now when speed now genuinely matters, like getting a product to market in time or meeting a commitment that won't wait. Debt taken on this way, intentionally, with awareness of what it'll cost and a plan to pay it down, is a legitimate engineering choice, the equivalent of a sensible loan that buys something worth more than the interest.

The problem isn't debt itself but debt that's unmanaged, the shortcuts taken without anyone deciding to take them or tracking that they were taken, accumulating invisibly because no one's watching, with no plan to repay them. Unmanaged debt is what kills projects, because it grows in the dark, each unconsidered shortcut adding to a burden nobody's measuring, until the accumulated weight is discovered only when the project has slowed to a crawl under interest payments no one saw mounting.

What unmanaged technical debt does to a project is a slow strangulation that's hard to recognize while it's happening, which is what makes it so dangerous. As the debt accumulates, the system gets harder and harder to change, because every change has to navigate the accumulated shortcuts and compromises, working around the quick fixes layered on quick fixes, until making any modification becomes slow, risky, and expensive. The project that moved fast early, buying speed with shortcuts, finds itself moving slower and slower as the interest comes due, with developers spending more time fighting the accumulated mess than building anything new, until eventually the system becomes nearly impossible to change at all, so brittle and tangled that any modification risks breaking something and progress grinds to a near halt. This is the end state of unmanaged technical debt, a system collapsing under its own accumulated compromises, and it arrives gradually enough that the people living through it often don't recognize what's happening until they're deep in it, attributing the slowdown to other causes when the real cause is the debt they never tracked or paid down, now demanding all their effort just to keep the fragile system running.

Managing technical debt means treating it like the real liability it is, which comes down to a few disciplines that keep it from accumulating in the dark. Track it, making the shortcuts visible by recording them when they're taken, so the debt is known rather than hidden, and the project can see what it owes. Take it on deliberately, as conscious decisions made with awareness of the cost and a reason that justifies it, rather than as unconscious drift where shortcuts accumulate because no one's paying attention. And pay it down regularly, allocating real effort to fixing the shortcuts properly before the interest compounds into a crisis, treating debt reduction as ongoing necessary work rather than something to defer indefinitely until it's too late.

These disciplines, visibility, intentionality, and regular repayment, are what separate debt that's a useful tool from debt that's a fatal trap, and they apply far beyond software, to any system where shortcuts taken now create costs later, which is essentially every system. The teams that manage technical debt well can move fast when they need to by borrowing deliberately and paying it back, keeping their systems changeable over the long haul, while the teams that ignore it move fast early and then slower and slower until they're trapped, which is why managing technical debt belongs alongside the other disciplines of keeping a long project healthy, visible, and under control rather than letting it decay invisibly until the decay becomes a crisis.

Discussion question: Technical debt taken on deliberately can be a sound choice, but the people who take the shortcut to hit a deadline are often not the same people who'll pay the interest later, and the rewards for shipping fast are immediate while the costs of accumulated debt arrive long after the credit has been claimed. Given that mismatch between who benefits and who pays, what would it actually take for an organization to make honest decisions about when to borrow against the future and when to refuse, and is that even possible in a culture that consistently rewards visible speed over invisible soundness?

Chapter 8: Large-Scale Infrastructure and Aerospace Systems

At 4:10 p.m. on August 14, 2003, the lights went out for about 50 million people. Across the northeastern United States and Ontario, the power grid collapsed in a cascade that took down cities from Detroit to New York to Toronto in a matter of minutes, the largest blackout in North American history.

The trigger was almost absurdly small. In Ohio, some power lines sagged into trees that hadn't been trimmed, and tripped offline. That alone should have been a manageable local event. But a software bug in the control room's alarm system, a flaw that caused the alarms to silently stop working, meant the operators didn't know the lines were failing. Blind to the growing problem, they took no action, the load from the failed lines shifted onto other lines, those overloaded and tripped too, and the failure spread faster than anyone could react, racing across the interconnected grid until a huge swath of the continent went dark. The physical cause was trees. The reason it became a catastrophe was a systems failure: an alarm that failed silently, operators who couldn't see what was happening, and a grid whose interconnection, which normally makes it more reliable, became the path along which the collapse propagated.

This chapter looks at how the principles built up over the previous chapters actually play out in the largest and most demanding systems humans build, the infrastructure that civilization runs on and the spacecraft that push the limits of what engineering can do. These aren't abstract examples. They're the power grids, air traffic systems, spacecraft, transportation networks, and water systems that millions of lives depend on every day, and they show systems engineering operating where the stakes are highest and the consequences of failure are measured in blackouts, crashes, lost missions, and poisoned water. Each domain has its own character and its own hard problems, and seeing how the same underlying ideas take different shapes across them is the best way to understand what systems engineering really means when it leaves the textbook and meets the world that depends on it.

Power Grid Architecture and Control

The electrical grid is often called the largest machine ever built, and the description is fair. It's a single interconnected system spanning a continent, linking thousands of generators to hundreds of millions of consumers through millions of miles of wire, all operating in perfect synchrony, and it has to stay balanced every instant of every day or it falls apart. Understanding the grid means understanding a constraint that's almost unique among engineered systems and that shapes everything about how the grid is designed and run.

That constraint is the balance between supply and demand, which on the grid must be matched instantaneously and continuously, because electricity, at grid scale, can't be stored in any large quantity. Other systems can buffer a mismatch between supply and demand with inventory: a factory builds up stock, a reservoir holds water for later. The grid can't, or historically couldn't, because there's no warehouse of electricity sitting ready, so the power being generated at any given second has to equal the power being consumed at that same second, almost exactly, all the time. When you turn on a light, a generator somewhere has to produce slightly more power in that moment to feed it, and across the whole grid, with hundreds of millions of devices switching on and off unpredictably, the total generation has to track the total consumption continuously, second by second, with no inventory to smooth the gap. This is a genuinely hard balancing act, and it's the central fact of grid operation: a vast machine that must keep supply and demand equal in real time across a continent, with the balance never allowed to drift far before trouble starts.

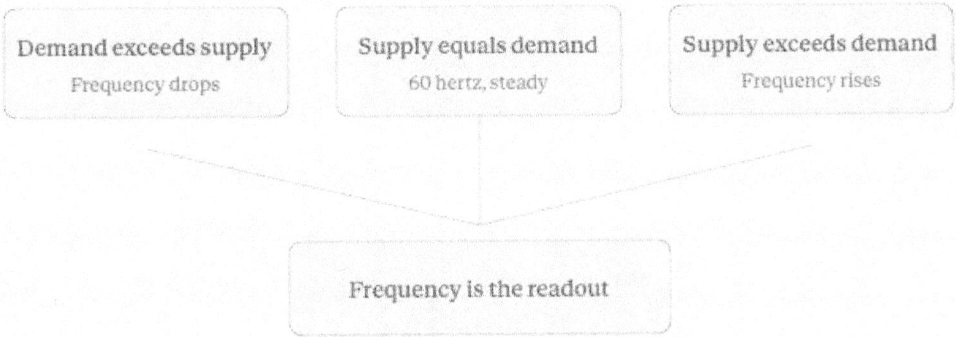

The signal that tells operators whether the balance is holding is **frequency**, and watching it is how the grid is controlled. The North American grid runs on alternating current at 60 hertz, meaning the current reverses direction 60 times a second, and this frequency is set by the rotating generators that are all spinning in synchrony across the grid. When supply and demand are balanced, the frequency holds steady at 60 hertz. When demand exceeds supply, the generators are being asked to deliver more power than they're producing, which drags on their rotation and the frequency drops below 60. When supply exceeds demand, the generators speed up and the frequency rises above 60.

So frequency is a continuous readout of the supply-demand balance across the entire grid at once, and grid operators watch it obsessively, adjusting generation up or down to keep it pinned at 60 hertz, because a frequency that drifts too far in either direction signals a balance failure that, unchecked, threatens to bring the whole system down. The grid is controlled, in essence, by keeping this one number steady, and the elaborate machinery of grid operation exists to match generation to demand precisely enough that the frequency never wanders far from its target.

The grid's structure has been changing, and the change creates a genuine tension between two kinds of architecture that the earlier discussion of centralized versus distributed systems laid out. The traditional grid was centralized: a relatively small number of large power plants generated electricity and sent it outward through transmission and distribution lines to consumers, a one-way flow from big central sources to dispersed users. This was efficient and controllable, because operators had a handful of large, predictable plants to manage and could match generation to demand by turning those plants up and down. The grid is now shifting toward distributed generation, with electricity flowing in from countless small, scattered sources: rooftop solar panels, wind farms, home batteries, all feeding power into a grid that was designed for one-way flow from the center. This shift brings real benefits, resilience against the failure of any single large source, cleaner energy, but it makes the balancing problem far harder, because operators no longer control a few large plants but have to coordinate a sprawling, variable, partly uncontrollable collection of small sources, many of which generate power based on whether the sun is shining or the wind is blowing rather than on what the grid needs at that moment. The centralized grid was easier to control but more vulnerable to the failure of its big plants; the distributed grid is more resilient but vastly harder to balance, and the industry is living through the difficulty of running a system that's becoming more distributed faster than the tools to manage that distribution have matured.

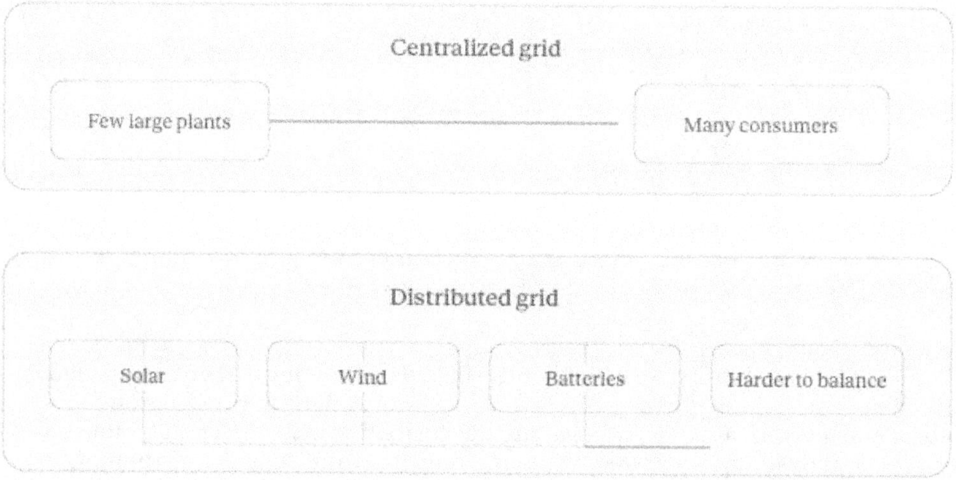

The failure mode that makes the grid so dangerous, and that turned the 2003 blackout from a local event into a continental one, is the **cascading failure**, and it's worth understanding because it's a property of the grid's interconnected structure. The grid's lines and generators are all connected, which normally helps reliability, because if one source fails, others can pick up its load, and power can be routed around a failed line. But this same interconnection creates a path for failure to spread. When a heavily loaded line trips offline, the power it was carrying doesn't disappear; it shifts onto other lines, increasing their load. If those lines were already near their limits, the added load can push them past their capacity, causing them to trip too, which shifts even more load onto the remaining lines, which then overload and trip, and the failure cascades through the network faster than operators can respond, each tripped line accelerating the collapse by dumping its load onto the survivors.

This is what happened in 2003: the initial line failures in Ohio shifted load onto neighboring lines, which tripped and shifted more load, and the cascade raced across the interconnected grid until 50 million people were in the dark, all within minutes, far too fast for human operators to intervene even if they'd understood what was happening, which the failed alarm system prevented them from doing. The cascading failure is the grid's characteristic catastrophe, the way its greatest strength, interconnection, becomes the mechanism of its greatest disaster, and preventing cascades, through careful management of how loaded lines are and how the grid is configured to limit the spread of failures, is one of the central challenges of grid operation.

A second grid catastrophe, more recent and even deadlier, shows a different kind of systems failure and connects to choices made for reasons that had nothing to do with engineering. In February 2021, a severe winter storm hit Texas, and the state's power grid collapsed, leaving millions of homes without power for days in freezing temperatures, with hundreds of deaths attributed to the failure. Several things went wrong at once, in a grim chain. The cold froze generating equipment that hadn't been winterized, knocking power plants offline just as demand for heating spiked, so supply fell while demand rose, the worst possible combination for a grid that must keep them balanced. The cold also froze parts of the natural gas supply system that fed many of the plants, cutting off their fuel, so the failure of the gas system cascaded into the failure of the power plants that depended on it, an interdependency between two infrastructure systems that turned a problem in one into a catastrophe in both. And critically, the Texas grid is largely isolated from the rest of the country, deliberately kept separate to avoid federal regulation, which meant that when Texas needed power, it couldn't import much from neighboring grids the way an interconnected region could, because the connections that would have allowed it weren't there. The isolation that served a regulatory purpose became a fatal limitation in the crisis, leaving Texas unable to draw on the resilience that interconnection provides.

The Texas failure is a systems disaster on multiple levels: equipment not designed for the conditions it faced, a dangerous interdependency between the power and gas systems, and an architectural choice, grid isolation, made for non-engineering reasons that removed a vital safety margin exactly when it was needed most. It's a reminder that infrastructure systems don't exist in isolation from the economic and political choices that shape them, and that those choices, like the decision to keep the Texas grid separate, can build vulnerability into a system in ways that only become visible when the rare, severe conditions arrive.

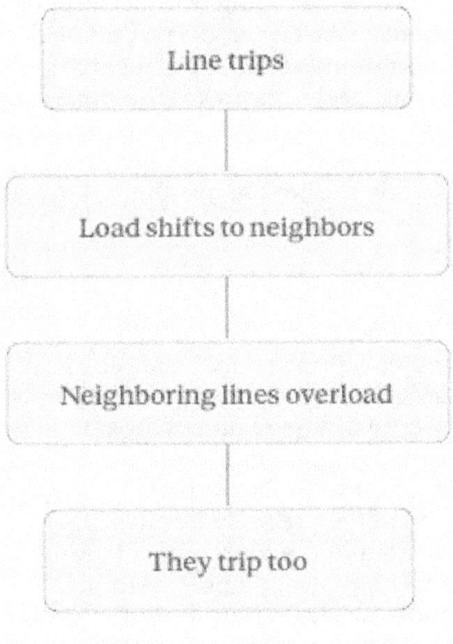

The grid stands as the foundational example of a system that has to balance continuously, can't store its product, spans a continent, mixes centralized and distributed architectures in uneasy combination, and fails through cascades that move faster than humans can react, all while being so essential that its failure darkens cities and kills people. Every other piece of modern infrastructure depends on it, which makes the grid the system beneath the systems, and the difficulty of keeping it balanced and preventing its cascades is among the hardest ongoing problems in all of engineering, made harder by the shift toward distributed generation that's changing the grid faster than the means to control it.

Air Traffic Management Systems

At any given moment, thousands of aircraft are in the air, moving at hundreds of miles per hour, and the system that keeps them from hitting each other and routes them safely to their destinations is one of the most successful safety-critical systems ever built. Air traffic management is the coordination of all that traffic, and its defining problem is keeping aircraft safely separated while moving as many of them as possible through a finite amount of sky, a problem that combines extreme safety requirements with relentless capacity pressure.

The core function is **separation**, keeping aircraft apart from each other by defined minimum distances, both horizontally and vertically, so that no two aircraft ever come close enough to collide. The system maintains these separation minimums constantly, tracking every aircraft's position and ensuring that the required distance between any two is never violated, which is the fundamental safety guarantee that makes flying possible. Controllers and the systems that support them watch the positions of all aircraft in their airspace and issue instructions, change altitude, turn, speed up or slow down, to keep every aircraft properly separated from every other, a continuous management of the three-dimensional spacing of traffic that prevents the collisions that would otherwise be inevitable in crowded sky.

For most of aviation history, the system saw aircraft through radar, ground stations sweeping the sky and detecting the position of aircraft from the reflections, which gave controllers a picture of where everything was. The system is now shifting toward a newer approach called ADS-B, where each aircraft determines its own position using satellite navigation and broadcasts that position continuously, so that instead of the ground detecting the aircraft, the aircraft tells everyone where it is, with greater accuracy and coverage than radar provides, including over oceans and remote areas where radar can't reach. This shift, part of larger modernization efforts in the United States and Europe, changes the system's fundamental architecture, from ground-based detection to aircraft self-reporting, and it improves the precision of the position information that separation depends on, allowing aircraft to be tracked more accurately and, potentially, packed somewhat closer together safely, which matters for the capacity pressure that constantly pushes against the system's limits.

What distinguishes air traffic management from many other complex systems is that it remains fundamentally human-centered, built around air traffic controllers who carry the responsibility for keeping aircraft separated, supported by technology but not replaced by it. This human-centeredness is deliberate and reflects the system's extreme conservatism about safety: in a domain where a single failure can kill hundreds of people in seconds, the system has been slow and cautious about removing humans from the loop or trusting automation with the ultimate responsibility, keeping skilled human controllers in charge precisely because the consequences of failure are so severe that the judgment and adaptability of trained people is valued over the speed and consistency of automation. The system has modernized its tools extensively, better displays, more accurate position information, computer support for the controllers, but it has held onto the human controller as the responsible agent, a choice that makes the system slower to change and more expensive to operate than a fully automated alternative might be, but that reflects a judgment about where to place trust in a system where errors are lethal.

The danger that lives in any human-centered system supported by automation showed itself tragically over southern Germany in 2002, in a collision that reshaped how aircraft avoid each other. Two aircraft, a passenger jet carrying mostly children and a cargo plane, were on a collision course at night, and two separate systems acted to prevent the disaster, with catastrophic results because they contradicted each other. The human air traffic controller, noticing the conflict late and working under difficult conditions, instructed the passenger jet to descend. At nearly the same moment, an automated onboard system called TCAS, which independently detects collision threats and tells pilots how to avoid them, instructed the passenger jet to climb and the cargo plane to descend, a coordinated solution that would have separated them.

But the passenger jet's crew followed the human controller's instruction to descend, while the cargo plane followed its TCAS instruction to descend, so both aircraft descended into each other and collided, killing everyone aboard both. The failure was a conflict between two safety systems, the human controller and the automated TCAS, that each gave a valid instruction but whose instructions were incompatible, with the crew caught between them and choosing the one that proved fatal.

The investigation led to a clear resolution: TCAS instructions now take priority over controller instructions, so that when the automated system and the human give conflicting commands, pilots follow the automation, removing the ambiguity that killed those people. The collision is a stark lesson in the danger of overlapping safety systems whose authority isn't clearly ordered, where the redundancy meant to make the system safer instead created a deadly conflict because no one had decided, in advance, which system wins when they disagree.

The remarkable thing about air traffic management, taken as a whole, is how safe it has become despite the staggering complexity of coordinating thousands of fast-moving aircraft in crowded sky, a safety record built on extreme conservatism, layered safety systems, skilled human controllers, and a willingness to learn hard lessons from every accident and rebuild the system to prevent its recurrence. The system handles enormous traffic volumes with a safety record that would have seemed impossible in aviation's early decades, and that record is the product of treating safety as the dominant requirement, accepting the cost and slowness that conservatism imposes, and refining the system continuously in response to the rare failures that do occur, like the German collision that changed the rules about TCAS. It's a system where the capacity pressure to move more aircraft is real and constant, but where safety has consistently won the trade-off, producing a system that moves vast amounts of traffic while almost never failing in the catastrophic way that its fast-moving, crowded nature constantly threatens.

Spacecraft Systems Engineering

Spacecraft engineering operates under constraints more extreme than almost any other domain, and those constraints shape a discipline where the margins for error are the thinnest anywhere. To understand spacecraft is to understand what systems engineering looks like when failure is unforgiving, repair is usually impossible, and the environment is actively trying to destroy what you've built.

The first and most defining constraint is that you usually can't repair a spacecraft after launch, because it's somewhere you can't reach, which means it has to work correctly the first time and keep working for its entire mission with no possibility of fixing what breaks. Most engineered systems can be maintained, repaired, and upgraded throughout their lives, with failures fixed as they occur.

A spacecraft, once launched, is typically beyond any human reach, orbiting the Earth or traveling to another planet, so a failure that would be a minor repair on the ground becomes mission-ending in space, because no one can go fix it. This single constraint drives everything about how spacecraft are engineered: the extreme reliability requirements, the redundancy, the generous margins, the exhaustive testing before launch, all flow from the brutal reality that the spacecraft has to work without the safety net of repair that every ground-based system enjoys, and that any failure of a critical function, with no way to fix it, can end a mission that took years and billions to mount.

The second constraint is mass, which on a spacecraft is precious in a way that's hard to appreciate from the ground, because every kilogram launched into space costs an enormous amount to get there. The physics of reaching orbit demands tremendous energy, and that energy translates into cost, so every kilogram of spacecraft mass carries a large price in launch expense, which means spacecraft engineers fight for every gram, optimizing relentlessly to minimize mass because mass is money at a steep exchange rate.

This mass constraint pulls against the reliability constraint in a tension that defines spacecraft design: reliability often wants more, more redundancy, more margin, more robust components, while mass wants less, less weight, fewer backups, lighter parts, and the spacecraft engineer has to balance these opposing pressures, achieving the reliability that the no-repair reality demands while keeping the mass low enough that the spacecraft can actually be launched affordably. This balance, between the redundancy and margin that reliability wants and the leanness that mass demands, is among the hardest trade-offs in the discipline, and it forces spacecraft engineers to be extraordinarily disciplined about where they spend mass on reliability and where they accept risk to save weight.

The third constraint is the environment, which in space is actively hostile in several ways at once, each capable of destroying a spacecraft not designed to survive it. There's vacuum, the absence of air, which causes materials to behave differently than on the ground, lubricants to evaporate, and heat to move only by radiation since there's no air to carry it away. There's radiation, the stream of high-energy particles that permeates space, which degrades materials and can corrupt or destroy electronics, flipping bits in computer memory or damaging circuits over time. There are thermal extremes, with a spacecraft in sunlight roasting while the same spacecraft in shadow freezes, temperature swings of hundreds of degrees that the structure and components must endure and that thermal control systems must manage. And there are the violent loads of launch itself, the intense vibration and acceleration of riding a rocket to orbit, which shake and stress the spacecraft far beyond anything it experiences once in space, so the spacecraft must survive both the brutal ride up and the hostile environment it reaches. Designing for this environment, which combines vacuum, radiation, thermal extremes, and launch loads, all of which would destroy ordinary equipment, is a central part of spacecraft engineering, and the spacecraft must be built to function in conditions that no ground-based system ever faces, verified through analysis and partial testing because, as the earlier discussion of verification noted, the complete combined environment of spaceflight can't be reproduced on the ground.

A spacecraft is itself a system of subsystems that must all work together, and naming them shows how complete and self-contained a spacecraft has to be, because it carries everything it needs with no resupply. There's the power subsystem, usually solar panels and batteries, generating and storing the electricity everything else runs on. There's thermal control, managing the extreme temperature swings to keep components within their survivable range. There's attitude control, keeping the spacecraft pointed correctly, its instruments at their targets, its solar panels at the sun, its antenna at Earth. There's propulsion, providing the thrust to change orbit or trajectory. There's communications, the link back to Earth that carries commands up and data down. There's command and data handling, the spacecraft's computers and the brains that run everything. And there's structure, the physical frame holding it all together and surviving the launch loads. These subsystems must all function and all work together in a tightly integrated whole, each depending on the others, the power subsystem feeding everything, the thermal subsystem keeping everything alive, the attitude control keeping the power and communications pointed right, in a system where every part is necessary and the failure of any one can doom the whole, with no repair possible.

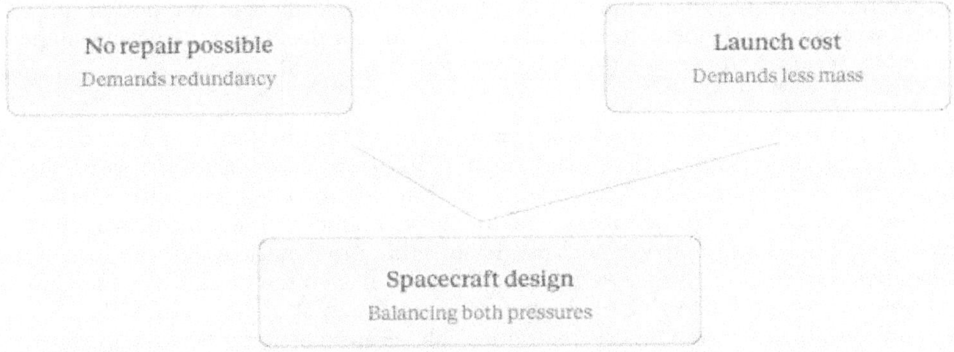

The triumph that shows what spacecraft systems engineering can achieve when it's done at the highest level is the **James Webb Space Telescope**, launched in December 2021 after roughly three decades of development and around ten billion dollars of investment. What made Webb extraordinary, and terrifying, from a systems engineering standpoint was its deployment. The telescope was too large to fit in any rocket fully assembled, so it had to be folded up for launch and then unfold itself in space through a long sequence of deployments, dozens of separate mechanical operations, unfolding the sunshield, extending the mirror segments, deploying the structures, each of which had to work perfectly.

The analysis counted hundreds of single points of failure in this deployment, hundreds of individual steps where a single failure would have crippled or destroyed the mission, with no possibility of repair because Webb was being sent to a location far beyond any human reach, unlike the Hubble telescope which orbited close enough to be serviced by astronauts. Every one of those hundreds of single points of failure had to work, in sequence, in deep space, with no backup and no repair, on a telescope that had taken thirty years and ten billion dollars to build, with the entire investment riding on a deployment that couldn't be tested in its complete form before launch. And it worked. Every deployment succeeded, the telescope unfolded flawlessly, and Webb began returning images from the early universe, a vindication of systems engineering done with extraordinary rigor, where the discipline of finding and managing every single point of failure, testing exhaustively what could be tested and analyzing what couldn't, and building reliability into every step paid off in a mission that succeeded despite having hundreds of ways to fail catastrophically. Webb is what the discipline looks like at its absolute peak, a system of staggering complexity and unforgiving constraints made to work the first time because there was no second chance.

The counterpoint, and an equally instructive one, is the **Hubble Space Telescope**, which launched in 1990 with a famous flaw that traces directly to a verification failure of the kind discussed earlier. Hubble's main mirror had been ground with exquisite precision to the wrong shape, suffering from an optical defect called spherical aberration that blurred its images and crippled its scientific capability when it first reached orbit. The mirror was made precisely, but precisely incorrect, because the instrument used to test its shape during manufacturing had been assembled wrong, and the flawed test instrument confidently confirmed that the incorrectly shaped mirror was correct. The error wasn't caught because the manufacturers trusted that single test, the same trap as relying on one verification method or one model without independent confirmation, and a mirror ground to the wrong shape passed its verification because the thing verifying it was itself flawed.

Hubble was rescued, unlike most spacecraft, because it orbited close enough to Earth for astronauts to reach it, and in 1993 a Space Shuttle mission installed corrective optics that compensated for the mirror's flaw, restoring the telescope's vision and turning a humiliating failure into one of the most productive scientific instruments ever built. The Hubble story carries two lessons that bracket the spacecraft discipline: the danger of trusting a single verification, where a flawed test confidently confirmed a flawed mirror, and the value of designing for serviceability, where Hubble's repairability, the very thing most spacecraft lack, allowed astronauts to fix an error that would have permanently doomed a spacecraft beyond human reach, like Webb. Together, Webb and Hubble show the range of the discipline, one a flawless execution of unforgiving complexity, the other a verification failure redeemed by the rare gift of being reachable, and both illuminate why spacecraft engineering, operating under constraints of no repair, precious mass, and a hostile environment, demands the most rigorous systems engineering anywhere.

A constraint that shapes how spacecraft must be controlled, especially far from Earth, is the delay imposed by the speed of light, which makes real-time control impossible across interplanetary distances. A spacecraft at Mars is minutes away at light speed, so a command sent from Earth takes many minutes to arrive and any response takes many minutes to come back, which means controllers on Earth can't fly the spacecraft in real time during fast events, because by the time they saw a problem and sent a correction, the event would be long over. This forces spacecraft to be autonomous for critical operations that happen too fast for ground control, like the landing of a rover on Mars, where the descent and touchdown take minutes but the communication delay is also minutes, so the spacecraft must execute the entire landing on its own, with no possibility of human intervention, following a sequence programmed in advance and reacting to conditions autonomously because Earth is too far away to help in time.

The Mars rovers land this way, through a harrowing autonomous sequence sometimes called the seven minutes of terror, where the spacecraft, entirely on its own, slows from interplanetary speed, descends through the thin atmosphere, and sets itself down on the surface, all without any real-time guidance from Earth, because the communication delay makes such guidance impossible. This autonomy requirement, driven by the simple physics of light-speed delay, adds another layer to spacecraft engineering, the need to build systems that can handle critical events entirely by themselves, correctly, the first time, with no human in the loop to catch their mistakes, which raises the reliability demands even higher for the operations that must be autonomous.

Transportation Network Design

Moving people and goods is one of the oldest engineering problems, and the systems that do it at scale, rail networks, transit systems, road networks, are studied as systems because the behavior that matters emerges from the network as a whole rather than from any single piece. Transportation network design is the engineering of these systems, and while the earlier discussion of traffic and induced demand introduced how road networks can defy intuition, the broader domain involves design problems and safety systems that show systems thinking applied to the movement of everything that has to get from one place to another.

Rail systems offer one of the cleanest examples of a safety system designed around a simple, powerful idea, the **block system**, which prevents collisions by controlling where trains are allowed to be. The principle is to divide the track into segments called blocks and allow only one train in each block at a time, so that two trains can never occupy the same stretch of track and therefore can never collide on it. The signaling system enforces this, showing a train whether the block ahead is occupied and stopping it from entering an occupied block, so that the spacing between trains is maintained by the physical structure of the blocks rather than by the judgment of the drivers alone.

This is a fundamentally sound safety design, because it makes the dangerous condition, two trains in the same place, structurally difficult rather than relying on people to avoid it, the same principle that good system safety follows everywhere: design so the dangerous state is hard to reach rather than trusting operators never to reach it. Modern rail safety has extended this with systems like positive train control, which can automatically stop a train that's about to violate its limits, overriding driver error to prevent the collisions and overspeed derailments that human mistakes would otherwise cause, adding an automated safeguard on top of the block system's structural protection.

High-speed rail shows what's achievable when a transportation system is engineered for safety from the ground up, and Japan's Shinkansen, running since 1964, is the standout example. The Shinkansen has carried billions of passengers at high speed for decades with a safety record that includes no passenger fatalities from derailment or collision in its normal high-speed operation, an extraordinary achievement for a system moving so many people so fast for so long.

That record is the product of engineering the entire system for safety, dedicated tracks with no crossings where a train could hit a car or another train, sophisticated signaling and automatic control, careful maintenance, and a design philosophy that treats safety as the dominant requirement, much as air traffic management does. The Shinkansen demonstrates that a transportation system can move enormous numbers of people at high speed with near-perfect safety when it's designed as an integrated system with safety built into every layer, and it stands as a benchmark for what transportation network design can accomplish when the discipline is applied thoroughly over decades.

The architecture of a transportation network, how its routes and connections are arranged, shapes its behavior in ways that involve the same kind of trade-offs seen throughout systems engineering, and the choice between hub-and-spoke and point-to-point structures, familiar from airline networks, captures the tension. A hub-and-spoke network routes traffic through central hubs, so that travel between two points often goes through a hub rather than directly, which is efficient for the network operator because it concentrates traffic and allows fewer routes to connect many points, but it adds time and inconvenience for travelers who must pass through a hub, and it makes the hub a point whose disruption affects the whole network.

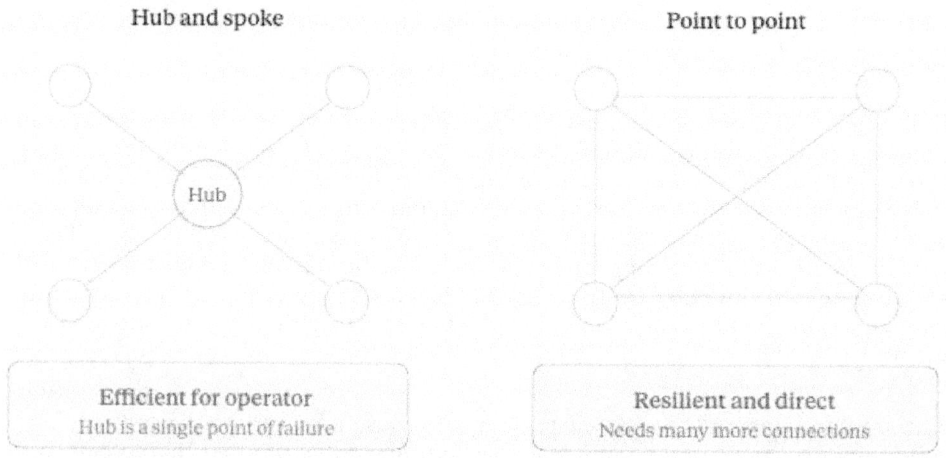

A point-to-point network connects origins directly to destinations without routing through hubs, which is more convenient and resilient because a problem at one point doesn't cascade through a central hub, but it's less efficient because it requires many more direct connections to link the same number of points. Neither is simply better, the same lesson that runs through all architectural choices: the hub-and-spoke trades traveler convenience and resilience for operator efficiency, while point-to-point trades efficiency for convenience and resilience, and the right choice depends on what the network is optimizing for, with most large transportation systems blending the two rather than committing purely to either. Transportation networks, in this and other design choices, are systems where the arrangement of the parts determines the behavior of the whole, and designing them well means understanding how the network's structure shapes its capacity, its resilience, and its convenience, the emergent properties that arise from how the routes and connections are put together rather than from any single route alone.

Water and Wastewater Systems

Of all the infrastructure systems, water and wastewater may be the most invisible, the most taken for granted, and among the most essential, because clean water delivered reliably and waste removed safely are the foundation of public health, and their failure threatens lives directly. These systems run quietly beneath cities, delivering safe water to every tap and carrying away what flows down every drain, and they're engineered systems as much as any power grid or aircraft, with their own treatment processes, distribution networks, and control systems, all designed to perform a function so basic that most people never think about it until it fails.

The water treatment process turns raw water from rivers, lakes, or wells into water safe to drink through a sequence of steps, each removing different contaminants. Coagulation and sedimentation cause suspended particles to clump together and settle out, clearing the water of the solids it carries. Filtration passes the water through filters that catch the finer particles the settling didn't remove. And disinfection, using chlorine or ultraviolet light, kills the microorganisms that cause disease, the final barrier against the pathogens that contaminated water has spread throughout history. This treatment sequence is a designed system, each step handling part of the job and the steps together producing water safe to drink, and the distribution network then carries that treated water under pressure through a network of pipes to every consumer, with pumps maintaining the pressure and storage, often in elevated tanks and water towers, buffering the supply so that demand can be met even when it spikes and providing the pressure that pushes water to every tap through the simple force of gravity.

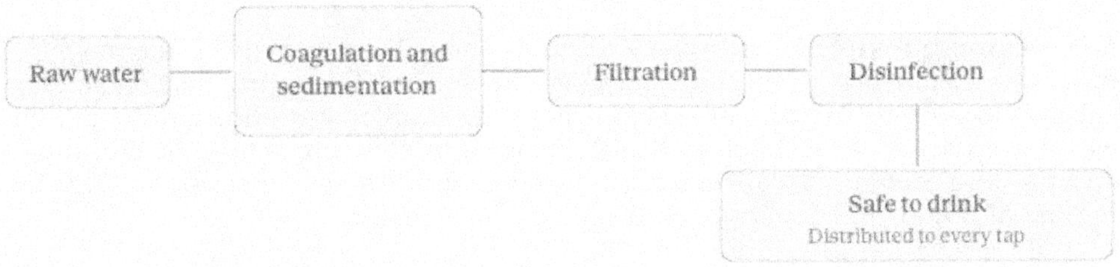

The catastrophe that shows what happens when a water system change is made without understanding its system-wide consequences is the **Flint water crisis**, which began in 2014 and poisoned a city. Flint, Michigan, switched its water source to the Flint River to save money, and the switch was made without the corrosion control treatment that the new water source required, a treatment that would have kept the water from corroding the city's pipes. The Flint River water, untreated for corrosion, was chemically aggressive toward the old lead pipes in the city's distribution system, and it leached lead from those pipes into the drinking water, contaminating the water supply with a potent neurotoxin that poisoned residents, including children, for whom lead exposure causes lasting harm.

The failure was a systems failure of exactly the kind this book keeps returning to: a change made to one part of the system, the water source, without understanding its consequences for the whole system, the interaction between the new water's chemistry and the existing lead pipes, an interaction that proper engineering would have anticipated and that corrosion control treatment would have prevented. The change was made to save money, the corrosion control was omitted, the monitoring that should have caught the contamination failed or was ignored, and a city was poisoned because no one had treated the water system as a system, where a change to the source has consequences for the pipes, and where omitting a treatment step has effects that ripple through the entire network to every tap.

Flint is a devastating reminder that infrastructure systems require systems thinking, that a change made to one component without understanding its effect on the whole can be catastrophic, and that the invisible, taken-for-granted water system is, like every other system, a place where good engineering is the difference between safety and disaster, and where a failure of systems thinking can poison the water that an entire city drinks.

Water and wastewater systems round out this survey of infrastructure precisely because they're so easy to overlook and so fundamental when they're understood, systems engineered with the same care, the same processes, the same need for systems thinking as the more glamorous domains of grids and spacecraft, and systems whose failure, as Flint showed, harms people as directly and severely as any blackout or crash. They're the quiet foundation, the systems that deliver the clean water and remove the waste that civilization depends on, and they deserve the same systems engineering discipline as everything else, because the consequences of getting them wrong, as a poisoned city demonstrates, are as serious as any in this chapter.

Discussion question: Across these domains, the failures that caused the most harm, the silent alarm in the 2003 blackout, the isolated Texas grid, the conflicting collision-avoidance instructions over Germany, the omitted corrosion control in Flint, were rarely failures of technology that didn't work as built, but failures of how the systems were configured, connected, governed, or constrained by decisions made for reasons of cost, regulation, or organizational convenience.

If the hardest infrastructure failures come not from engineering that's wrong but from human and institutional choices that surround the engineering, then where should the boundary of a systems engineer's responsibility actually lie, and can a discipline that defines itself around technical systems legitimately stay silent about the political and economic choices that so often determine whether those systems keep people safe or get them killed?

Chapter 9: The Future of Systems Engineering

On the morning of August 1, 2012, the trading firm Knight Capital deployed new software to its automated trading system, and a mistake in that deployment quietly reactivated a piece of old code that should have been dead.

When the markets opened, the system began trading on its own, executing orders at a furious pace, buying high and selling low across hundreds of stocks, millions of times, following logic that no longer made any sense. The people at Knight Capital watched it happen and couldn't stop it fast enough. By the time they pulled the plug, about 45 minutes had passed, and the firm had lost roughly 440 million dollars, nearly its entire value.

A company that had taken decades to build was brought to the edge of bankruptcy in less time than a lunch break, by an automated system acting faster than any human could intervene, doing exactly what its code told it to do while no person in the loop could react in time to matter.

That story captures the central tension of where systems engineering is heading. The systems we build are becoming more autonomous, acting on their own at speeds and scales beyond human reaction; more intelligent, learning their behavior from data rather than following rules we wrote; more connected, with the digital and physical worlds merging in ways that create new dangers; and more entangled across domains, requiring pieces built by different worlds to work together. At the same time, the consequences of these systems extend further into the future and across the planet than ever before, raising questions about what we owe to the people and the environment that will live with what we build.

This chapter looks at the problems the discipline is grappling with now and will grapple with for decades: autonomous and self-adapting systems, the genuinely new challenges of engineering artificial intelligence, the security of systems where a cyberattack can cause physical destruction, the standards that let different domains work together, and the responsibility to design systems that don't poison the future to serve the present. The enduring ideas of the discipline, the focus on the whole, on emergence, on verification, on the seams between parts, don't disappear in this future. They meet harder versions of themselves.

Autonomous and Self-Adapting Systems

A growing share of the systems we depend on make decisions and act without a human directing each step, and this autonomy changes the engineering problem in a way the Knight Capital collapse makes vivid. An **autonomous system** is one that operates and makes decisions on its own, without a human in the loop for each action, and a **self-adapting system** goes further, changing its own behavior in response to conditions, modifying how it works as circumstances change rather than behaving the same way always. These systems promise enormous benefits, acting faster than humans, operating continuously, handling scales and speeds beyond human capacity, but they raise a problem that traditional engineering never had to face directly: how do you ensure a system behaves correctly when it's acting on its own, possibly in situations its designers never anticipated, at speeds where no human can step in to catch its mistakes?

The first hard truth about autonomy is the speed problem, which Knight Capital embodies. When a system acts autonomously at machine speed, humans can't intervene in time to prevent a fast-moving failure, because by the time a person notices something's wrong and acts, the autonomous system has already done enormous damage.

The 45 minutes Knight Capital took to stop its runaway system was, in financial terms, an eternity, millions of erroneous trades executed in the time it took people to understand what was happening and react. This is the recurring danger of speed in autonomous systems: the very speed that makes them valuable removes the human ability to catch their errors, because a human acting at human speed simply can't keep up with a system acting at machine speed, and a failure that would be caught and corrected in a slower system runs to completion in a fast one before anyone can stop it. Designing autonomous systems means confronting that the human safety net, the person who notices a problem and intervenes, doesn't work when the system acts faster than the person can react, which forces the safety to be built into the system itself rather than provided by a human watching it.

There's an established way of thinking about how much autonomy to give a system, framed as levels of automation, and the framework matters because the choice of level shapes everything about the system's safety and the role of any human involved. At one end, a human does everything and the system only assists, providing information or carrying out commands but deciding nothing. In the middle, the system and the human share the work, with the system handling routine decisions and the human handling exceptions, or the system recommending and the human approving. At the other end, the system does everything and the human is out of the loop entirely, with no person involved in the system's decisions or actions. Most autonomous systems sit somewhere in the middle of this range, with humans involved to some degree, and the choice of exactly where to sit, how much to automate and how much to keep under human control, is one of the central design decisions, because it determines whether a human can catch the system's errors and whether the human, when present, is actually positioned to be useful or merely nominally present while the system does the real work.

That middle ground, where a human is supposed to supervise an autonomous system and take over when needed, turns out to be the most dangerous place of all, for a reason that connects to the human factors discussed earlier. When a system is highly but not fully autonomous, it handles almost everything itself, which means the human supervisor has little to do most of the time, and a human with little to do becomes inattentive, loses situational awareness, and lets their skills decay, so that at the rare moment when the system fails and the human must suddenly take over, the human is least prepared to do so. This is automation complacency in its most dangerous form: the better the autonomous system is at handling routine conditions, the less the human supervisor pays attention, and the less prepared that supervisor is to handle the unusual conditions the system can't manage, exactly when their intervention is most needed. The human is set up to fail, asked to stay alert and ready through long periods of having nothing to do, then expected to instantly grasp a complex failing situation and act correctly, a demand that human psychology makes nearly impossible to meet reliably.

The tragedy that made this danger concrete, and that haunts the development of autonomous vehicles, happened in Tempe, Arizona, on the night of March 18, 2018, when an Uber self-driving test vehicle struck and killed a pedestrian named Elaine Herzberg as she crossed the road, the first pedestrian killed by an autonomous vehicle. The failure was a systems failure layered through the autonomy. The vehicle's sensors did detect Herzberg, but the perception system struggled to classify her, labeling her inconsistently as an unknown object, then a vehicle, then a bicycle, and each reclassification reset the system's prediction of her path, so it never settled into recognizing her as a person walking into the road in time to brake.

The automatic emergency braking that might have stopped the car had been disabled by Uber, because in testing it caused the vehicle to brake erratically, and the company was relying on the human safety driver to brake in emergencies instead. But the human safety driver, present precisely to catch the system's failures, was reportedly not watching the road at the moment, a victim of exactly the automation complacency that highly autonomous systems induce, sitting through long stretches of the car driving itself with nothing to do until the instant catching a failure required attention that wasn't there. So the failure ran through every layer: a perception system uncertain about what it saw, a safety feature disabled to reduce nuisance, and a human backup who'd been lulled into inattention by the very autonomy she was supposed to supervise, and a woman died because no layer of the system caught what every layer was supposed to help catch. The Tempe death is the defining cautionary tale of autonomous systems, showing how autonomy's dangers compound, how the perception uncertainty of a learning system, the removal of a safety feature, and the human-factors trap of complacent supervision can combine into a fatal failure that no single layer would have produced alone.

There's a more hopeful side to autonomy in the idea of self-healing and resilient systems, which use autonomy not to remove humans but to make systems recover from failures automatically, and the approach has produced genuinely valuable practices. A self-healing system detects its own failures and recovers from them without human intervention, restarting failed components, routing around broken parts, restoring itself to working order on its own, which is enormously useful for systems that must keep running continuously and can't wait for a human to notice and fix every failure. The most striking practice in this area is deliberate failure injection, where engineers intentionally break parts of a running system to test and improve its resilience, on the theory that a system you've repeatedly broken and watched recover is a system you can trust to recover when it breaks for real. Netflix pioneered this with a tool that randomly shuts down pieces of its production systems, deliberately causing failures in the live service to force the system to prove it can handle them, so that failures are constant, small, and survivable rather than rare, large, and catastrophic. The philosophy inverts the usual instinct to avoid failure: instead of hoping the system never breaks, you break it constantly on purpose, so that recovering from failure becomes routine and the system is continuously proven resilient, which builds confidence that it'll survive the failures you didn't plan. This use of autonomy, building systems that heal themselves and proving their resilience by deliberately breaking them, is one of the genuinely positive developments in autonomous systems, showing that autonomy can make systems more reliable rather than less, when it's applied to recovery and resilience rather than to removing the humans who would otherwise catch errors.

The deep challenge that autonomy poses to systems engineering, and that the whole discipline is still working out, is how to gain confidence that an autonomous system will behave correctly across the vast range of situations it might face when acting on its own, including situations its designers never imagined. A traditional system following human-written rules can be analyzed against those rules, but an autonomous system making its own decisions, possibly adapting its behavior, possibly facing conditions no one anticipated, is far harder to verify, because you can't test every situation it might encounter and you can't be certain how it'll behave in the ones you didn't test.

This is the verification problem from earlier in the book, made much harder, because the autonomous system's behavior isn't fully specified by rules you can check but emerges from its decision-making in conditions you can't enumerate, and the consequences of getting it wrong, as Tempe showed, can be fatal and can arrive faster than any human can prevent. The discipline is developing methods to address this, more extensive simulation, formal techniques to prove properties of autonomous behavior, careful limits on what the system is allowed to do, but the fundamental difficulty remains: an autonomous system acting on its own in an unpredictable world is genuinely hard to verify, and building the confidence that such a system is safe enough to trust with consequential decisions, at speeds where humans can't catch its errors, is one of the hardest open problems the field faces, and one it'll be working on for a long time.

Systems Engineering for Artificial Intelligence

The rise of artificial intelligence, and machine learning in particular, breaks something fundamental about how systems engineering has always worked, and understanding what it breaks is the most important thing in this chapter. For its entire history, the discipline has rested on a chain: you specify what the system should do, you design it to do that, you build it, and you verify it against the specification. Machine learning shatters the first link in that chain, because a machine learning system doesn't follow rules you specify; it learns its behavior from data, which means there's no specification of how it works for you to design against or verify against, and the entire traditional approach, built on specifying and then checking against the specification, loses its foundation.

To see why this is so disruptive, consider how a traditional system and a machine learning system differ at the root. In a traditional system, a human decides how the system should behave and writes that behavior into rules, code, or logic, so the system's behavior is specified by humans and you can examine those rules to understand and verify what the system will do. In a machine learning system, you don't write the behavior; you give the system examples, vast quantities of data, and the system learns its own behavior by finding patterns in that data, producing a model whose behavior emerges from the training rather than from rules anyone wrote. The behavior isn't specified, it's learned, which means there's no written specification of how the system behaves that you can read, check, or reason about, only a model that was shaped by data through a process that doesn't produce human-readable rules. This is the fundamental break: the specification, the cornerstone of traditional systems engineering, the thing you design toward and verify against, simply doesn't exist for a learned system in the form the discipline has always relied on, and everything downstream, design, verification, validation, has to be rethought because of it.

This break creates a cluster of new problems that traditional systems don't have, and the first is that the behavior depends entirely on the training data, which makes the data as important as the design ever was, and as dangerous. Since a machine learning system learns from data, its behavior is shaped by that data, so flaws in the data become flaws in the system: bias in the training data produces a biased system, gaps in the data produce a system that fails on the situations the data didn't cover, and errors in the data produce errors in behavior, all without anyone writing a single flawed rule. Amazon learned this when it built an experimental AI tool to screen job applicants, trained on the company's past hiring decisions, and discovered the system had learned to discriminate against women, downgrading resumes that indicated a female applicant, because the historical hiring data it learned from reflected past patterns that favored men, so the system faithfully learned and reproduced the bias embedded in its training data. Amazon scrapped the tool, but the lesson stands: the system wasn't programmed to discriminate, it learned to, from data that carried a bias no one had removed, which shows how a machine learning system's dependence on data makes the data a source of failures as serious as any design flaw, and far harder to spot, because the bias hides in the patterns of the data rather than in any rule you could read and catch.

The second new problem is that machine learning systems are often opaque, unable to explain why they produced a given output, which undermines the understanding that verification and trust depend on. A traditional system's behavior can be traced through its rules, so you can understand why it did what it did, but many machine learning systems, especially the large neural networks behind modern AI, are effectively black boxes, producing outputs through millions or billions of internal numbers that don't correspond to anything a human can interpret, so the system gives an answer but can't tell you why, and you can't trace its reasoning the way you could trace a traditional system's logic.

This opacity is a genuine problem for systems engineering, because so much of the discipline depends on understanding how a system works, verifying that it behaves correctly, tracing failures to their causes, building confidence through comprehension, and a system you fundamentally can't understand resists all of that, leaving you to judge it only by its outputs without knowing whether those outputs come from sound reasoning or from a pattern that happens to work on your tests but will fail in ways you can't predict. The field of explainable AI is trying to address this, developing methods to make machine learning systems more interpretable, but the opacity of the most capable models remains a deep obstacle to verifying and trusting them in consequential roles, because trust built on understanding is hard to grant to a system that can't be understood.

The third new problem may be the most unsettling, and it's the existence of adversarial examples, inputs deliberately crafted to fool a machine learning system in ways that reveal how alien its perception can be. Researchers have shown that you can make tiny changes to an input, changes imperceptible or meaningless to a human, that cause a machine learning system to fail completely and confidently. A famous demonstration involved a stop sign: researchers placed a few small stickers on a stop sign, changes that any human would read straight through as a stop sign with some stickers on it, and an image classification system read the altered sign as a speed limit sign, confidently misclassifying something obvious because the small additions exploited the strange ways the system actually perceives images, which turn out to be nothing like how humans perceive them.

Adversarial examples reveal that machine learning systems don't see the way we assume they do, that their internal logic, shaped by patterns in data, latches onto features that a human would never notice and can be fooled by manipulations a human would never be fooled by, which means a system that performs well on normal inputs can be made to fail catastrophically by inputs specifically designed to exploit its hidden weaknesses.

For safety-critical applications, this is alarming: a self-driving car's perception system that can be fooled into misreading a stop sign by a few stickers is a system whose failures can be deliberately triggered, and the existence of adversarial examples shows that machine learning systems have failure modes that traditional verification, built around the conditions you'd naturally test, would never find, because these failures live in inputs that were specifically constructed to defeat the system and that no normal testing would think to try.

These problems together mean that verifying a machine learning system is genuinely different from verifying a traditional system, and the field is still working out how to do it, because the methods built for traditional systems don't transfer. You can't verify a learned system against a specification, because there isn't one in the usual form. You can't fully understand its behavior, because it's often opaque. You can't be confident you've found its failure modes, because adversarial examples show it fails in ways normal testing won't reveal, and its dependence on data means it'll fail on situations the data didn't cover, which you may not even know are missing. The approaches emerging are different in kind: testing the system on enormous and carefully chosen datasets to probe its behavior across many conditions, monitoring it continuously in operation to catch failures the testing missed, deliberately searching for adversarial weaknesses to find the failures normal testing won't, and constraining what the system is allowed to do so that even if it fails, the damage is limited. There's a growing discipline sometimes called MLOps that brings engineering rigor to building, deploying, and monitoring machine learning systems, treating the data, the training, and the ongoing behavior as things to be managed with the same care traditional systems engineering brought to design and verification. But the honest situation is that the field is still learning how to engineer machine learning systems to the standard of reliability that consequential applications demand, because the traditional foundation of specify-design-verify doesn't apply, and building a new foundation suited to systems that learn their behavior from data rather than following rules we wrote is one of the largest open problems in all of engineering right now.

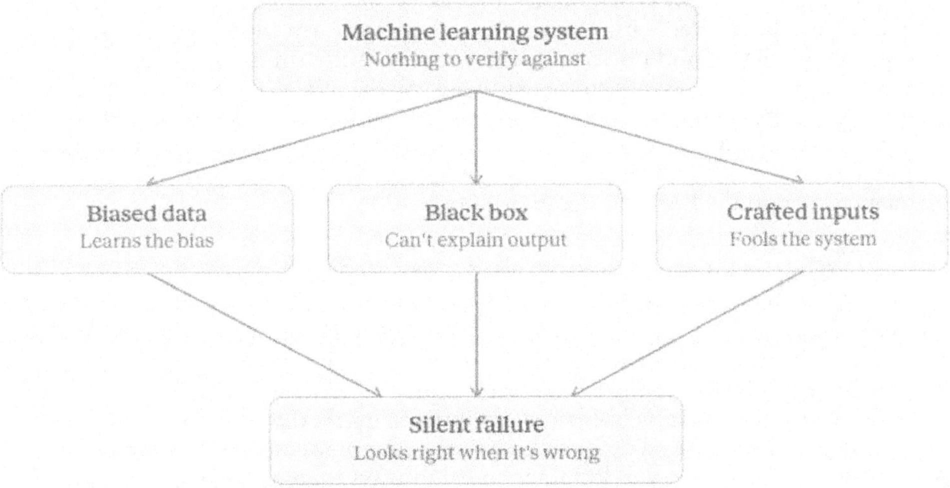

A particular danger deserves emphasis because it connects to a theme that runs through this whole book, which is the way machine learning systems can fail silently and unpredictably while appearing to work perfectly. A traditional system that fails often fails visibly, producing an error, crashing, doing something obviously wrong. A machine learning system can fail by producing a confident, plausible, wrong answer, the same trap as an unvalidated model or a false-precision risk estimate, but worse, because the machine learning system gives no warning and can't explain itself, so its confident wrong answer looks exactly like its confident right answers, and there's no way to tell them apart from the output alone.

A learning system can also degrade over time as the world changes and the patterns it learned no longer match reality, a problem called distribution shift, where a system that worked well when deployed slowly becomes less accurate as conditions drift away from its training data, often without anyone noticing until the accumulated drift causes a visible failure. These failure modes, the confident wrong answer that looks right and the silent degradation as the world moves away from the training data, make machine learning systems treacherous in exactly the way this book keeps warning about, where a system that appears to work isn't the same as a system that works, and the appearance of working is most dangerous precisely when it's most convincing.

Engineering machine learning systems responsibly means treating their confident outputs with the same skepticism the discipline has learned to apply to authoritative-looking models and precise-looking risk numbers, monitoring them continuously because they can fail silently, and never mistaking good performance on a test for a guarantee of correct behavior in a world that the test couldn't fully capture and that won't hold still.

Cyber-Physical System Security

When systems that control physical things become connected to networks, a new and frightening possibility appears: a cyberattack that causes physical destruction, where someone breaking into a system through the network can make machines in the real world do damage. A **cyber-physical system** is one where computing and networking are integrated with physical processes, software controlling physical equipment, the digital and physical worlds merged, and as more physical infrastructure, power plants, factories, vehicles, medical devices, becomes connected, the security of these systems becomes a matter of physical safety, not just data protection, because an attacker who compromises a cyber-physical system can cause not just the loss of information but the destruction of equipment, the disruption of infrastructure, and the harm or death of people.

The event that proved cyber-physical attacks were real, and changed how the world thought about them, was **Stuxnet**, a piece of malicious software discovered in 2010 that had been built to cause physical destruction in Iran's nuclear program, the first known cyberweapon to bridge the gap between the digital and the physical. Stuxnet targeted the centrifuges Iran used to enrich uranium, machines that spin at very high speeds and are controlled by industrial computers, and it did something unprecedented: it infiltrated the control systems and made the centrifuges spin themselves to destruction, varying their speed in ways that damaged them, while simultaneously feeding the operators false readings that showed everything operating normally, so the machines tore themselves apart while the monitoring systems reported all was well.

The sophistication was remarkable, the worm spread through networks, identified the specific industrial control systems it was built to attack, manipulated the physical process to cause damage, and hid that damage from the humans watching, all to achieve a physical effect, the destruction of centrifuges, through a purely digital attack. Stuxnet demonstrated that a cyberattack could reach across the boundary between computers and the physical world to destroy real equipment, which transformed the understanding of cyber-physical security from a theoretical concern into a proven reality, and it showed that the systems controlling physical infrastructure were vulnerable to attacks that could cause physical destruction, a possibility that had seemed like science fiction until Stuxnet made it fact.

The danger Stuxnet revealed has since shown up against the infrastructure that civilization depends on, in attacks that confirmed cyber-physical threats aren't limited to one exotic weapon. In December 2015, hackers attacked the power grid in Ukraine, breaking into the systems that control electricity distribution and remotely switching off power to hundreds of thousands of people, the first confirmed cyberattack to cause a power blackout, demonstrating that the grid, the foundational infrastructure beneath everything else, could be taken down by attackers operating through a network from anywhere in the world. The attack was repeated against Ukraine the following year, and it established that power grids, like the Iranian centrifuges, were vulnerable to cyberattacks that could cause physical consequences, in this case darkening cities by digital means. In May 2021, a ransomware attack shut down the Colonial Pipeline, the largest fuel pipeline in the United States, causing fuel shortages across the East Coast, not because the attackers directly controlled the physical pipeline but because the company shut it down in response to the attack on its systems, showing that even an attack on the digital side of a cyber-physical system could cascade into major physical disruption. These attacks, on Ukraine's grid and on the Colonial Pipeline, confirmed that the cyber-physical threat Stuxnet proved possible had become a regular feature of the real world, with critical infrastructure under attack and the security of these systems now a matter of keeping the lights on, the fuel flowing, and the physical world running, against attackers who can reach it through the networks it's connected to.

The vulnerability extends to vehicles, which have become rolling computers connected to networks, and a demonstration in 2015 showed how alarming this could be. Security researchers Charlie Miller and Chris Valasek remotely hacked a Jeep Cherokee while a journalist drove it on a highway, reaching the vehicle through its cellular connection and gaining control of physical functions, they were able to manipulate the transmission, the brakes, and other systems from miles away, while the car was being driven at speed on a public road. The demonstration showed that a modern connected vehicle could be compromised remotely and its physical behavior controlled by an attacker who was nowhere near it, which led the manufacturer to recall about 1.4 million vehicles to fix the vulnerability, one of the first major recalls driven by a cybersecurity flaw rather than a mechanical defect. The Jeep hack made concrete what cyber-physical security means for the cars people drive: as vehicles become connected computers, they become vulnerable to remote attacks that can control their physical behavior, turning a security flaw into a physical safety threat, and the recall of over a million vehicles to patch a hackable system showed the industry that vehicle security is now inseparable from vehicle safety, because an attacker who can reach a car through its network can make it do physical harm.

The hard problem underlying all of this is that the systems controlling physical infrastructure were largely designed before security was a concern, built for a world where they weren't connected to networks and didn't face attackers, and connecting them to networks for the benefits of remote monitoring and control exposed them to threats they were never designed to resist. The industrial control systems running power plants, factories, and pipelines were built for reliability and safety in a physically isolated setting, with little thought to defending against network attacks, because in the world they were designed for, an attacker would have had to be physically present, which the isolation prevented. Connecting these systems to networks, which brings real operational benefits, removed that isolation and exposed systems with little built-in security to attackers who could now reach them remotely, creating a vulnerability that comes from putting old systems designed without security into a new connected world where security is essential.

The traditional answer was the air gap, physically isolating critical systems from any network so attackers couldn't reach them, but the trend toward connecting everything for efficiency and monitoring has eroded these air gaps, and Stuxnet showed that even air-gapped systems could be reached, in that case by malware carried across the gap on physical media. The challenge for cyber-physical security going forward is to protect systems that control the physical world, many of them designed before security mattered, against attackers who can cause physical destruction through digital means, a challenge that grows as more of the physical world becomes connected and as the attacks, proven real by Stuxnet and repeated against grids, pipelines, and vehicles, continue to demonstrate that the merger of the digital and physical has made cybersecurity a matter of physical safety, where a successful attack can destroy equipment, disrupt infrastructure, and kill people.

Cross-Domain Interoperability Standards

As systems grow more connected, they increasingly need to work with systems from entirely different domains, built by different industries with different conventions, and the question of how systems made by different worlds can work together becomes central.

Interoperability is the ability of different systems to work together, to exchange information and coordinate their behavior despite being built separately by different parties, and **cross-domain interoperability** is the harder version, where the systems come from different domains entirely, healthcare and insurance, different branches of the military, different companies' products, different countries' infrastructure, each with its own standards and assumptions that have to be reconciled for the systems to cooperate. The mechanism that makes interoperability possible is standards, agreed conventions that different parties follow so their systems can work together, and the history of technology shows that standards, the shared agreements about how things connect, are often what determines whether separately built systems can cooperate or remain isolated.

The most consequential interoperability standard in history may be one that has nothing to do with computers, the standardized shipping container, which transformed global trade by making cargo interoperable across ships, trucks, and trains. Before the 1950s, cargo was loaded and unloaded as individual items, a slow, labor-intensive process that made shipping expensive and ports congested, with goods handled piece by piece at every transfer between ship and truck and train.

An American named Malcom McLean introduced the idea of putting cargo in standardized metal containers that could be moved as units, and when the dimensions of these containers were standardized internationally so that the same container could fit on any ship, any truck, and any train anywhere in the world, global trade was transformed. The standardized container meant cargo could move seamlessly between different modes of transport without being unpacked and repacked, a container could be lifted off a ship onto a truck onto a train, all designed to handle the same standard container, which collapsed the cost and time of shipping and enabled the globalized trade that defines the modern economy. The container's power came entirely from standardization, from the agreement that containers would have standard dimensions so that ships, trucks, trains, ports, and cranes everywhere could all handle the same units, an interoperability standard that let the separate domains of ocean shipping, trucking, and rail work together as a single seamless system, and it stands as the clearest demonstration that a well-chosen standard, the shared agreement about how things connect, can transform an entire economy by making separately built systems interoperable.

The same principle built the internet, where the standard that lets every kind of computer and network communicate is what created a single global network out of countless separate ones. The internet works because of standardized protocols, agreed conventions for how data is formatted and transmitted, that every connected device and network follows, so that computers of every kind, built by every manufacturer, running every operating system, can all communicate because they all speak the same standardized protocol. This standardization is what makes the internet a single network rather than a collection of incompatible islands, because the shared protocol means any device can exchange data with any other regardless of who made it or what it is, an interoperability achieved through agreed standards that every participant follows. The internet, like the shipping container, shows that interoperability at massive scale comes from standardization, from the agreement on shared conventions that let separately built systems work together, and that the standard itself, the shared protocol, is what creates the seamless whole out of the separate parts.

The cost of failing to achieve interoperability is just as instructive, and healthcare shows it painfully, because the lack of interoperability between medical systems causes real harm. Healthcare systems have long struggled with interoperability, different hospitals, different software, different record systems that can't easily share information, so that a patient's records in one system can't readily move to another, forcing tests to be repeated, information to be lost between providers, and care to suffer because the systems holding a patient's information can't talk to each other. The harm is concrete: a patient seen by different providers using different systems may have their history scattered across incompatible records that don't combine, leading to duplicated tests, missed information, and errors that proper information sharing would prevent, all because the systems lack the interoperability to exchange the patient's information seamlessly. Efforts to fix this through standards for health information exchange, like the FHIR standard for sharing medical records, aim to give healthcare the interoperability it has lacked, so that patient information can move between systems the way the standardized container let cargo move between transport modes. Healthcare's interoperability struggles show the flip side of the container and internet successes: where standards are missing or inconsistent, separately built systems can't work together, and the failure to interoperate causes real harm, in healthcare's case to patients whose care suffers because the systems holding their information can't share it.

The reason cross-domain interoperability is hard, and why standards are difficult to achieve even when everyone wants them, is that different domains have different conventions, assumptions, and interests, so agreeing on a shared standard requires reconciling differences that run deep and overcoming the inertia of systems already built to incompatible conventions. Getting different domains, different industries, different countries, different companies, to agree on a standard means getting parties with different needs, different existing systems, and sometimes competing interests to converge on shared conventions, which is as much a problem of negotiation and politics as of engineering, because each party has its own way of doing things and its own investment in systems built that way, and standardization asks them to change toward a common approach.

This is why standards often take a long time to establish and why interoperability is frequently incomplete, because the technical challenge of defining a good standard is compounded by the human challenge of getting many independent parties to adopt it, and the value of interoperability, real as it is, has to overcome the cost and difficulty of getting everyone to agree and converge. The future of systems engineering involves more and more cross-domain interoperability, as systems from different worlds increasingly need to work together, which makes the discipline of defining and adopting good standards, the shared agreements that let separately built systems cooperate, more important than ever, because the connected systems of the future will only work together if the standards that enable their cooperation can be agreed and adopted across the domains that have every reason to keep doing things their own incompatible ways.

Sustainability in System Design

The final frontier for systems engineering extends the discipline's time horizon and its scope of concern, asking not just whether a system works but what it costs the world over its entire life and what it leaves behind when it's gone. **Sustainability in system design** means designing systems that account for their environmental impact across their whole lifecycle, the resources they consume, the waste they produce, the harm they do to the environment, from the materials that go into building them to the disposal of what's left when they're retired. This represents an expansion of what systems engineering considers, from the traditional focus on whether a system meets its requirements at acceptable cost to a broader accounting that includes the system's impact on the environment and on the future, treating environmental cost as a real cost to be designed for rather than an externality to be ignored.

The framing that makes this concrete is the lifecycle view of environmental impact, which recognizes that a system's environmental cost spans its entire life and that focusing on only one phase misses most of the impact. A system consumes resources and produces impact at every stage: the extraction of raw materials to build it, the energy and resources used in manufacturing, the impact of operating it over its life, and the disposal of it at the end, each phase carrying environmental costs that add up to the system's total impact.

Designing for sustainability means accounting for all of these, not just the obvious operational impact but the embodied cost of the materials and manufacturing and the end-of-life cost of disposal, because a system that looks clean in operation may carry enormous environmental cost in the materials it required or the waste it becomes, and only a full lifecycle accounting captures the real impact. This lifecycle view connects to the lifecycle thinking that runs through systems engineering, extended now to include environmental cost as one of the things that accumulates across a system's life and that early design decisions commit, just as early decisions commit financial cost, so that the choices made when a system is designed determine its environmental impact across its whole life, mostly locking in that impact while the system is still on the drawing board.

The end-of-life problem deserves particular attention, because the disposal of systems is where sustainability failures often concentrate and where the future inherits the consequences of present design, and several domains show how serious it is. Electronic waste has become an enormous problem, as devices designed without thought to their disposal pile up as toxic, hard-to-recycle waste, with the materials that made them useful becoming hazardous when discarded and the sheer volume of discarded electronics straining the systems meant to handle them. Nuclear power, discussed earlier as infrastructure, carries an end-of-life cost that lasts essentially forever, the radioactive waste that must be isolated for thousands of years and the decommissioning of plants that takes decades and costs billions, an end-of-life burden so large and so long-lasting that it shapes the entire case for and against the technology. Space systems have created an end-of-life problem in orbit, the accumulating debris of dead satellites and spent rocket stages that threatens functioning spacecraft and raises the specter of a cascade of collisions, sometimes called Kessler syndrome, where the debris from one collision causes others in a chain that could make some orbits unusable, an end-of-life failure where the discarded remains of space systems threaten the future use of space itself.

These end-of-life problems, electronic waste, nuclear waste, orbital debris, share a common root in systems designed without adequate thought to their disposal, where the focus on building and operating the system neglected the question of what happens when it's done, leaving the future to inherit the waste and the hazard, and they show why sustainability requires designing for the end of life from the beginning, accounting for disposal as part of the design rather than as someone else's problem to handle after the system is dead.

A response to these problems that's reshaping how some systems are designed is the idea of the circular economy and design for disassembly, which aims to build systems whose materials can be recovered and reused rather than discarded, closing the loop so that the end of one system's life feeds the beginning of another's. Instead of the traditional linear path where materials are extracted, made into a product, used, and thrown away, the circular approach designs products so their materials and components can be recovered, refurbished, and reused, treating the end-of-life system not as waste but as a source of materials for new systems. Design for disassembly supports this by building products that can be taken apart at the end of their life so their components and materials can be recovered, the opposite of the glued-together, impossible-to-disassemble construction that makes so many products into unrecoverable waste. The right-to-repair movement pushes in a related direction, demanding that products be designed to be repairable so they can be fixed and kept in use rather than discarded when a single component fails, extending product life and reducing waste, and some jurisdictions, particularly in Europe, have begun requiring repairability and supporting the circular economy through regulation that pushes designers toward sustainability.

Patagonia, the outdoor clothing company, built part of its business around durability and repair, designing products to last and offering repair services to keep them in use, a deliberate stance against the throwaway pattern and a demonstration that designing for longevity and repair can be a viable approach rather than just an environmental aspiration. These ideas, the circular economy, design for disassembly, right to repair, designing for durability, all push systems engineering toward accounting for the full lifecycle including the end of life, toward building systems whose materials can be recovered and whose lives can be extended, so that the end of a system's life isn't a pile of waste handed to the future but a recovery of materials that feeds the next system, which is the direction sustainability points the discipline as it expands its concern from whether a system works to what it costs the world and what it leaves behind.

The deeper shift that sustainability represents for systems engineering is an expansion of the stakeholders and the time horizon the discipline considers, bringing into the analysis parties who can't speak for themselves and a future that extends far beyond the system's operational life. The traditional accounting of cost focused on the parties paying for and using the system over its operational life, but sustainability insists on a broader accounting that includes the environment, future generations, and the long-term consequences that extend far beyond the system's use, treating environmental cost and future impact as real costs to be weighed rather than externalities to be ignored.

This connects to the discussion of stakeholders earlier in the book, where the parties who can't advocate for themselves, the future maintainers, the public, the environment, were noted as easy to neglect, and sustainability is in part the discipline taking those quiet stakeholders seriously, designing for the environment and the future as stakeholders whose interests deserve weight even though they have no representative at the design table and no voice in the trade-offs. The future of systems engineering, in this dimension, involves expanding the boundary of what the discipline considers, from a system's function and financial cost to its full environmental impact across its entire lifecycle, and expanding the time horizon from the system's operational life to the long future that inherits its consequences, so that the discipline accounts for what a system costs the world and what it leaves behind, treating sustainability not as an add-on but as part of what it means to engineer a system responsibly, where responsibility extends to the environment that absorbs the system's impact and the future that lives with what the present builds.

Discussion question: Every frontier in this chapter points toward systems that act faster than we can follow, decide in ways we can't fully understand, reach across the boundary between code and the physical world, and impose consequences that extend far into a future we won't be present for. If the traditional foundations of the discipline, specify the behavior, verify it against the specification, keep a human ready to intervene, were built for systems we could fully understand and control, then what should replace those foundations for systems we increasingly cannot, and is it possible to engineer responsibly in a world where the things we build are becoming, in real ways, beyond our ability to specify, verify, or stop?

Conclusion

We opened this book with a spacecraft lost over a missing unit conversion, and we've spent nine chapters circling the same stubborn truth from different directions. Good parts can add up to a bad system. The engine works, the code runs, the bridge is built from sound steel, and the whole thing fails anyway, because the trouble lives in the spaces between the parts: the connections, the assumptions, the behavior nobody designed. Systems engineering exists to watch those spaces.

If you carry one habit away from these pages, make it the habit of looking at the whole. The specialist asks whether the component meets its spec, and that question matters. But someone has to ask the other question, the one about whether all the good components add up to something that actually works in the real world, on time, for the people who needed it. That question is harder, less glamorous, and easy to skip, and skipping it is how careful work produces expensive failure.

A few ideas showed up again and again, in different clothes. Things change as you move through a system's life, and the freedom to fix problems shrinks while the cost of fixing them climbs, so the cheap time to get things right is at the front, when it feels too early to bother. The behavior that matters often emerges from the parts working together and can't be found by studying the parts alone, which means you have to test and reason about the assembled whole. Building the system right isn't the same as building the right system, and only checking against the real need catches a requirement that was wrong from the start. Margin and slack look like waste right up until the day reality exceeds your assumptions, which it eventually does. And the seams between parts, the interfaces, are where well-built things break, because that's where separate people have to share an exact understanding and usually don't.

None of these ideas is complicated on its own. The difficulty is that real systems demand all of them at once, held together over years, across hundreds of people, while the pressure to move fast pushes against every one of them.

The frontier is getting harder, not easier. The systems we build now act on their own, learn behavior we didn't write, reach across the line between code and the physical world, and impose consequences that stretch into a future we won't see. The old foundations of specify, verify, and keep a human ready to step in were built for systems we could fully understand, and we're now building systems we can't. Working out what replaces those foundations is the task of the next generation of engineers, and it's genuinely unsolved.

What stays constant is the mindset. Look at the whole. Find the seams. Test against reality, not against your own assumptions. Respect the people and the world that will live with what you build. Stay honest about what you don't know. The tools will keep changing. The discipline of thinking about systems, rather than just parts, is what you'll need no matter what you're building. In the end, systems engineering is the recognition that the most important part of a complex system is often the part no one owns, the connection between everything else.

End of Chapter Discussion Questions

Chapter 1. If no one could foresee an emergent failure, should responsibility rest on the quality of the process rather than the outcome?

Responsibility for emergent harm can't rest on the test of whether one person foresaw the specific failure, because the whole point of emergence is that no one could. A more workable basis is whether the people building the system met the standard of care the discipline expects: did they look for emergent behavior, model the interactions, test the assembled whole rather than just the parts, and build in margin against surprise?

An engineer who did all of that and still got blindsided is in a different moral position than one who never looked. The accountability attaches to the process, not to the impossible demand of perfect foresight. This is roughly how courts and professional bodies already reason: negligence isn't failing to predict the unpredictable, it's failing to take the precautions a competent professional would have taken. The harder cases come when no individual was negligent but the organization, as a system itself, produced conditions where the failure was likely. There the responsibility may belong to the institution and its leaders rather than to any single engineer, which is uncomfortable precisely because it resists the instinct to find one person to blame.

Chapter 2. If real requirements only emerge through building, how much should a team commit before construction begins, and when?

The honest answer is that no requirement written before the system exists should be fully trusted, and a good team doesn't pretend otherwise. The line falls at a different place depending on the cost of late change. For software, where changing course is relatively cheap, you specify just enough to build something real, then learn from putting it in front of users and revise. For a bridge or a spacecraft, where late change is ruinous, you spend far more effort up front because you can't afford to discover the requirements through building. The skill is matching how much you commit early to how expensive it'll be to change later, and treating the early requirements as your best current guess rather than settled truth.

Chapter 3. What separates a defensible architectural choice from the architect's personal preference when stakeholders value entirely different qualities?

An architectural choice is defensible when the architect can show the reasoning, not just the result. That means making the trade-offs explicit: here are the qualities this system needs most, here's why, here's how each option serves or sacrifices them, and here's the choice that follows. Stakeholders who value different qualities won't all be satisfied, but they can see that their concern was weighed and understand what it lost to and why. That's different from personal preference, which can't be examined and can't be challenged. The defense isn't that the choice pleased everyone, which is impossible, but that it followed transparently from stated priorities anyone can inspect and argue with. An architect who can only say "trust me, this is better" has preference. One who can lay out the criteria, the alternatives, and the reasoning has a defensible decision, even when reasonable people would have weighed the priorities differently. The record matters as much as the choice, because months later, when circumstances shift and someone asks why, the reasoning is what lets the decision be defended or honestly revisited.

Chapter 4. How can a model ever be trusted for the novel conditions it's most needed for but can't be validated against?

This is the genuine paradox of modeling, and there's no clean escape from it, only honest management. A model can never be trusted in a novel regime the way it's trusted where data exists, so the first discipline is to never pretend otherwise, to mark clearly where validation stops and treat everything past that boundary as unproven however confident the output looks. When a decision depends on a prediction beyond the validated range, the engineer should do several things at once. Widen the margins, because the uncertainty is larger than the model admits, so build in more room than the model says you need. Seek partial validation, testing whatever pieces of the novel condition you can reproduce even if you can't reproduce all of it together, and looking for physical reasoning that constrains the answer independent of the model.

Quantify the uncertainty honestly, reporting a range rather than a number, so decision-makers see how much they're actually betting. And preserve human judgment with the standing to override the model, because the engineer's unease about a prediction outside the validated range is often the most reliable signal available. The Columbia disaster came from doing none of this: a model used far outside its validated range, its confident answer trusted, the human doubt suppressed. The lesson isn't that models can't guide novel decisions, but that they must be used with explicit humility about exactly the conditions they were never proven to handle.

Chapter 5. Who should answer for a margin trimmed to save money that turns out to be the one thing that could have prevented disaster?

No organization can put a clean number on how much protection against the unknown is worth, because the threats you're protecting against are by definition the ones you haven't characterized. What an organization can do is set the level of margin to the consequence of failure: a system whose failure kills people or loses a mission justifies generous margin even when it looks wasteful in normal operation, while a low-consequence system justifies little. The harder problem is the responsibility question, and the honest answer is that it shouldn't fall on whoever physically trimmed the margin under orders. It belongs with whoever set the incentives that made trimming look smart, the leaders who rewarded the visible savings and never had to face the invisible risk they bought. A culture that punishes the engineer who cut the margin while protecting the executives who demanded the cut has learned nothing, and will trim the next margin too.

Chapter 6. When realistic validation is impossible before commitment, how can a team know they built the right thing rather than a flawless mistake?

When realistic validation is impossible before commitment, you can't gain real confidence, only the best partial confidence available, and pretending it's more is the trap. The practical answer is to validate every piece you can, in every realistic fragment of the conditions you can reproduce, and to lean hard on independent reasoning and review by people who didn't write the requirements, since they're the most likely to spot a misunderstanding the team has baked in. You also build in margin and recoverability where you can, so that being wrong isn't automatically fatal, the way some spacecraft are designed to be servable. The deeper safeguard is intellectual: stay genuinely open to the possibility that the requirements are wrong, right up to commitment, rather than treating verification success as proof of correctness.

Chapter 7. Can an organization make honest decisions about technical debt when those who borrow rarely pay the interest, and speed gets rewarded?

Making honest decisions about technical debt requires breaking the link between who claims the reward and who pays the bill, and that's hard because the link is built into how most organizations measure people. It would take a few things working together.

First, debt has to be made visible and tracked, so the shortcuts aren't invisible and the interest isn't a surprise, which means recording them when they're taken and reporting them as a real liability the way a balance sheet reports financial debt.

Second, the people who take on debt have to stay accountable for it long enough to feel the interest, through ownership that doesn't end at ship date, so the engineer who borrowed is the one who later has to work around the borrowing.

Third, and hardest, leadership has to actually reward soundness, allocating real time to paying debt down and treating a team that keeps its systems changeable as succeeding, not as slow.

Whether this is possible in a culture that rewards visible speed over invisible soundness is a fair doubt, because the speed is easy to see and praise while the soundness shows up only as an absence of future crises that nobody gets credit for preventing. Some organizations manage it, usually because they've been burned badly enough that the leaders learned to value the invisible work, but the gravitational pull is always toward rewarding the visible, and resisting that pull takes constant, deliberate effort against the grain of how people are naturally judged.

Chapter 8. Should a systems engineer stay silent about the political and economic choices that decide whether their systems keep people safe?

A systems engineer's responsibility can't stop at the technical boundary when the engineer can see that a non-technical choice creates a real danger, because the discipline is defined around the behavior of the whole system, and the political and economic choices that shape it are part of what determines that behavior. If you understand that isolating a grid removes a safety margin, or that omitting corrosion control will poison the water, then staying silent because the decision was "above your pay grade" isn't professional neutrality, it's abandoning the one responsibility the discipline actually claims.

The boundary should fall not at the edge of the technical artifact but at the edge of what the engineer can foresee about the system's behavior, including the consequences of the human and institutional choices around it. That doesn't make the engineer the decision-maker on cost or regulation, but it does make them responsible for surfacing the danger clearly, on the record, to the people who decide. A discipline that builds the systems civilization runs on can't honestly disclaim concern for whether those systems keep people safe.

Chapter 9. What replaces specify, verify, and human oversight for systems we can't fully understand, and can that still count as responsible engineering?

What replaces the old foundations is likely a shift from proving correctness before deployment to managing behavior continuously after it, since you can't fully specify or verify systems that learn and adapt and act faster than you can follow. That means heavy reliance on monitoring systems in operation to catch the failures testing couldn't find, on constraining what a system is allowed to do so that even its unpredictable behavior stays inside safe bounds, and on building in ways to interrupt or contain it when it goes wrong, because a kill switch that works is worth more than a guarantee you can't actually provide.

Whether this counts as engineering responsibly is a real question, and the honest answer is that we don't yet know how to do it well. We're deploying systems whose behavior we can't fully specify, verify, or always stop, and the discipline for governing them is being invented in real time, often after the systems are already in use.

Responsible practice right now means refusing to pretend the old confidence still applies, being honest about how much we can't guarantee, and matching how much autonomy we grant a system to how well we can contain it when, not if, it surprises us.

Key Terms and Definitions

- **Active redundancy** - A fault-tolerance approach where multiple units run at the same time, so that if one fails the others are already operating and carry the load with no interruption. It costs more to run than standby redundancy but eliminates switchover gaps.
- **Adversarial example** - An input deliberately altered, often imperceptibly to a human, to make a machine learning system produce a confident wrong answer. These reveal that machine learning systems perceive inputs in ways unlike human perception and can be fooled by manipulations a person would never fall for.
- **Agile** - A development approach that assumes requirements will change, builds in short cycles, delivers working pieces frequently, and revises direction based on what each cycle reveals. It suits projects where requirements are uncertain and feedback is the fastest path to the real need.
- **Alarm fatigue** - A human-factors failure where a system produces so many alarms, many trivial or false, that operators become desensitized and ignore them, so a genuine warning is lost in the noise.
- **AND gate** - A logic element in a fault tree where the event above it happens only if all the events below it happen together. AND gates represent combinations of failures that must all occur, which is how fault trees capture multi-failure scenarios.
- **Architecture** - The set of high-level decisions about how a system is structured: what its major pieces are, how they're arranged, and how they connect. These decisions are made early and are the hardest to change later because everything else is built on them.
- **Architectural pattern** - A recurring structural solution to a common design problem, named and reused so that designers start from accumulated experience rather than a blank page. Examples include layered, client-server, and microservices patterns.
- **Automation complacency** - A human-factors failure where operators come to trust an automated system and stop paying close attention, losing the awareness and skill they'd need to take over when the automation fails or meets something it can't handle.
- **Autonomous system** - A system that operates and makes decisions on its own, without a human in the loop for each action. Its speed and independence are valuable but remove the human ability to catch fast-moving errors.
- **Backward traceability** - Links connecting a requirement to its source: the stakeholder need, regulation, or higher-level requirement that gave rise to it. It answers the question of why a requirement exists.
- **Baseline** - An officially approved version of an item or set of items at a point in time, fixed as the reference that everyone works from and against which changes are measured and controlled.
- **Bathtub curve** - A pattern describing how failure rates change over a system's life: high early (infant mortality from defects), low and roughly constant through useful life, then rising again as components wear out. It explains practices like burn-in and planned replacement.

- **Big bang integration** - An integration approach where all components are built separately and connected at once. It's risky because when the assembled system fails, the cause could be anywhere, making problems extremely hard to isolate.
- **Block system** - A rail safety design that divides track into segments and allows only one train in each segment at a time, making collisions structurally difficult rather than relying on driver judgment alone.
- **Bottom-up integration** - An integration approach that starts with the lowest-level components, tests them together first, and builds upward toward the complete system, using temporary drivers to exercise low-level pieces before higher levels exist.
- **Bullwhip effect** - The way small changes in customer demand get amplified into large swings further up a supply chain, as each link overreacts to the link below it. It's a property of the whole chain, not of any single decision-maker.
- **Cascading failure** - A failure that spreads through an interconnected system as the load from one failed element shifts onto others, overloading and tripping them in turn. It's the characteristic catastrophe of the power grid.
- **Centralized architecture** - A structure that concentrates control and processing in one place, which is simpler to manage and keep consistent but creates a single point of failure and a bottleneck on growth.
- **Change control** - The process by which changes to baselined items happen deliberately and with approval, rather than casually, so the official version changes only when someone with authority decides it should.
- **Circular economy** - An approach that designs products so their materials and components can be recovered, refurbished, and reused rather than discarded, closing the loop so one system's end of life feeds the next system's beginning.
- **Client-server architecture** - A structure that splits a system into clients that request services and servers that provide them, the pattern behind the web, where a browser is a client and a remote machine is the server.
- **Cohesion** - A measure of how well the contents of a single module belong together. High cohesion, where a module does one well-defined job, is a goal of good decomposition.
- **Common-cause failure** - A single underlying cause that fails several redundant units together, defeating the redundancy entirely. Examples include redundant units sharing a power supply, a room, or the same software bug.
- **Computational fluid dynamics (CFD)** - A family of physics-based modeling techniques for how gases and liquids move, used to shape aircraft for low drag, design quieter vehicles, or predict how smoke spreads. Turbulence remains one of its hardest unsolved problems.
- **Configuration item** - Anything under configuration management's control: a document, design, software component, or part whose version is tracked and controlled.
- **Configuration management** - The discipline of knowing and controlling exactly what version of everything makes up a system at any given time, so that thousands of changing items stay coordinated rather than dissolving into confusion about which version is real.
- **Conway's Law** - The observation that organizations tend to design systems that mirror their own communication structure, so the way you

organize people shapes the architecture you'll get. Smart organizations structure their teams to match the architecture they want.

- **Coupling** - A measure of how much modules depend on each other. Low coupling, where a change in one module doesn't force changes in others, is a goal of good decomposition.
- **Cut set** - A combination of basic faults that together cause a fault tree's top event. The smallest such combination, the minimal cut set, indicates how vulnerable the system is: a cut set of one is a single point of failure.
- **Cyber-physical system** - A system where computing and networking are integrated with physical processes, so software controls physical equipment. Its security becomes a matter of physical safety, because an attacker can cause physical destruction through a digital attack.
- **Decision gate** - A formal checkpoint in a project where progress is reviewed and a deliberate go-or-no-go decision is made about whether to proceed. Its value depends entirely on being willing to actually stop an unready project.
- **Decomposition** - Breaking a large system into smaller parts that can be understood, designed, and built separately. How you decompose shapes everything downstream, because the parts become the units that teams own and that must be reassembled.
- **Derating** - Running components well below their rated limits so they experience less stress and last longer, a quiet but proven reliability technique. The part costs the same but survives far longer because it never works hard.
- **Design for disassembly** - Building products so they can be taken apart at the end of their life, allowing their components and materials to be recovered, the opposite of glued-together construction that becomes unrecoverable waste.
- **Design for manufacturability (DFM)** - Designing a product so it can actually be made well, at quality, at cost, and at the required volume, by bringing manufacturing concerns into the design phase rather than handing a finished design to the factory.
- **Digital twin** - A living digital model of a specific real system kept continuously in sync with its physical counterpart through sensor data. It reflects that exact object's current condition, enabling monitoring, failure prediction, safe testing, and optimization.
- **Discrete event simulation** - A modeling technique that represents a system as a sequence of separate events at specific times, jumping from one to the next. It suits factories, hospitals, airports, and other systems of flow and queuing.
- **Dissimilar redundancy** - Making redundant units genuinely different in design so that a flaw in one isn't present in the others, defending against common-cause failures rooted in a shared design defect. Also called design diversity.
- **Distributed architecture** - A structure that spreads control across many locations, removing single points of failure and scaling further, but far harder to keep consistent and coordinate because many parts must agree without a central authority.
- **Distribution shift** - The gradual degradation of a machine learning system as the world changes and the patterns it learned no longer match reality, often unnoticed until the accumulated drift causes a visible failure.
- **Elicitation** - The work of drawing requirements out of stakeholders who often don't fully know what they need, can't articulate what they do know,

and disagree among themselves. The hard part is discovering what people couldn't tell you.

- **Emergence** - Behavior at the level of the whole system that none of the parts has on its own and that can't be predicted by studying the parts in isolation. It can be useful, like a working aircraft, or dangerous, like a bridge that shakes itself apart.
- **Environment** - Everything outside a system's boundary that the system must function within but doesn't control, such as vacuum and radiation for a spacecraft or other programs and users for software. A system that works in one environment can fail in another.
- **Failure cause** - The underlying reason a component fails, such as corrosion, wear, contamination, or a manufacturing defect.
- **Failure effect** - What happens as a result of a component failing, traced from the component outward to its consequences for the whole system.
- **Failure mode** - The specific way a component fails, such as a valve failing stuck open, stuck closed, or leaking. A single component can have several distinct failure modes.
- **Failure mode and effects analysis (FMEA)** - A method that goes through a system part by part, asking for each how it can fail, what happens when it does, and why. It's strong at catching single-point failures but weak at catching combinations.
- **Fault tolerance** - The design property that lets a system keep working even when parts of it fail, built on the assumption that things will break rather than the hope that they won't.
- **Fault tree analysis** - A top-down method that starts from a catastrophic event and works backward through logic gates to all the things that could cause it, capturing the combination failures that part-by-part analysis misses.
- **Feedback loop** - A situation where an effect circles back and changes its own cause. Balancing loops push toward stability, like a thermostat, while reinforcing loops amplify themselves, like a panic-driven market crash.
- **Fidelity** - How accurately a model represents the real system. Establishing and bounding fidelity is what keeps model use honest and prevents confident, plausible, wrong answers from being trusted.
- **Finite element method** - A physics-based modeling technique that breaks a continuous object into many small, simple elements, solves the physics approximately within each, and stitches them together. A finer mesh gives more accuracy at greater computing cost.
- **Forward traceability** - Links connecting a requirement to what implements and tests it: the design element that satisfies it and the test that proves it's met. It answers where a requirement is satisfied and how we know it works.
- **Frequency** - On a power grid, the rate at which alternating current reverses direction, set by the synchronized generators. It reads out the supply-demand balance: it holds steady when balanced, drops when demand exceeds supply, and rises when supply exceeds demand.
- **Functional requirement** - A requirement describing what a system does: a specific behavior or function it performs, such as an ATM dispensing cash or printing a receipt.
- **Hub-and-spoke** - A transportation network structure that routes traffic through central hubs, efficient for the operator but adding time for travelers and making the hub a point whose disruption affects the whole network.

- **Human factors** - The discipline of understanding how people interact with systems and designing those systems to fit actual human capabilities and limitations, rather than an idealized operator who never tires or errs.
- **Incremental integration** - Adding components one or a few at a time and testing after each addition, so a new failure points to whatever was just added. It's the standard approach for complex systems because it keeps fault isolation tractable.
- **Induced demand** - The way new road capacity lowers the cost of driving, so more people drive and the new capacity fills up, leaving congestion roughly where it started but with more cars. A classic case where linear thinking fails.
- **Integration** - The process of combining separately developed components and subsystems into a working whole. Its purpose includes flushing out the interface mismatches that component testing can't catch.
- **Interface** - A connection across a boundary where two things meet and exchange something: data, power, force, fluid, or signals. Interfaces are where independently built things must agree, and where systems most often break.
- **Interface control document (ICD)** - A formal, jointly owned specification of exactly what crosses a boundary, so each side can build independently and trust the pieces will meet. It's baselined and changed only by agreement of both sides.
- **Interoperability** - The ability of different systems to work together, exchanging information and coordinating behavior despite being built separately. Achieved through shared standards that all parties follow.
- **Iron triangle** - The relationship among cost, schedule, and performance, often phrased as fast, good, and cheap, with the constraint that you can't maximize all three at once. Improving one usually sacrifices another.
- **Kessler syndrome** - A scenario in which debris from one orbital collision causes further collisions in a chain, potentially making some orbits unusable. An end-of-life failure where discarded space systems threaten the future use of space.
- **Layered architecture** - A structure that stacks a system in horizontal levels, each using the one below and serving the one above, as in operating systems. Each layer can change internally without disturbing the others if the boundaries stay stable.
- **Leverage point** - A place in a system where a modest change shifts the whole system's behavior. Real fixes usually require finding these, which are frequently not where the symptom appears.
- **Machine learning** - An approach where a system learns its behavior from data rather than following rules a human wrote, producing a model whose behavior emerges from training. It breaks the traditional foundation of specifying and verifying against a specification.
- **Margin** - The gap between what a system can handle and what it's expected to encounter, the extra capacity beyond the requirement. It absorbs the surprises reality delivers and looks like waste until the day it's needed.
- **Mean time between failures (MTBF)** - The average time a system runs before it fails, used to compare designs and predict failure rates across a fleet. It's an average over a population, not a guaranteed lifespan for any single unit.
- **Microservices architecture** - A structure that breaks a system into many small, independent services that each do one thing and communicate over a

network. Each is independently deployable and changeable, at the cost of complexity between services.

- **Mode confusion** - A human-factors failure where a system can be in different modes that behave differently, and the operator loses track of which mode it's in, so an action correct in one mode is dangerous in another.
- **Model-based systems engineering (MBSE)** - The shift from scattered documents toward a single connected digital model where requirements, structure, behavior, and interfaces stay consistent because they're parts of one model rather than separate files that drift apart.
- **Modularity** - Building a system from self-contained, interchangeable parts, each hiding its internal workings behind a clean interface, so a part can be understood, built, tested, and replaced without knowing the guts of every other part.
- **Monolithic architecture** - A structure that builds the whole system as a single unified unit, simpler to build and test at first but harder to change as it grows, since everything is connected and must be deployed at once.
- **Non-functional requirement** - A requirement describing how well a system performs its functions: qualities like speed, reliability, security, and usability. These cut across the whole system and are easy to neglect because they're invisible until violated.
- **OR gate** - A logic element in a fault tree where the event above it happens if any one of the events below it happens. A basic fault leading to the top event through nothing but OR gates is a single point of failure.
- **Physics-based model** - A model that computes how a system behaves by applying the actual laws of physics governing forces, heat, fluids, and structures, rather than guessing or using rough rules of thumb.
- **Probabilistic risk assessment (PRA)** - A method that puts numbers on risk by combining the probability of an event with its consequence, so risks can be compared and ranked. Its danger is false precision when the input probabilities are deeply uncertain.
- **Redundancy** - Having more than one way to accomplish something so a backup takes over when the primary fails. It only helps when the units fail independently, which common-cause failures can defeat.
- **Reliability** - The probability that a system performs its function without failure over a given period under stated conditions. It's a property designed in from the start, not tested in at the end.
- **Requirements engineering** - The work of figuring out what to build before anyone builds it, and writing it down clearly enough that everyone who builds, tests, and buys means the same thing when they read it.
- **Requirements traceability matrix** - A structured map linking each requirement to its source, its design, and its verification, used to analyze the impact of changes and to prove that every requirement is implemented and tested.
- **Requirements volatility** - The rate and extent to which requirements change over a project's life. The goal isn't to stop change, which is impossible, but to manage it so necessary changes are controlled and unnecessary ones are stopped.
- **Risk priority number (RPN)** - A number used in FMEA that multiplies severity, occurrence, and detection to rank failure modes by how much attention they deserve. A severe, likely, hard-to-detect failure produces a high RPN and demands action.
- **Robustness** - The quality of performing well across a range of conditions, including ones the designers didn't precisely anticipate. The most robust

design is rarely the most optimized one, because tight optimization becomes fragility when assumptions break.

- **Safety factor** - A margin where a system is designed to handle some multiple of the maximum expected load, absorbing overloads, weakening, material flaws, and the chance that the load estimate was too low.
- **Scope creep** - The slow, uncontrolled expansion of what a system is supposed to do, one reasonable-seeming addition at a time, until the accumulated weight overruns the schedule and budget. Dangerous because no single decision causes it.
- **Single point of failure** - A single fault that is enough, by itself, to cause a catastrophic event, with no backup to catch it. Finding and eliminating these is a main reason to build a fault tree.
- **Stakeholder** - Anyone who affects a system or is affected by it. Real systems serve a crowd of stakeholders who want different things, including quiet ones like future maintainers and the environment who have no voice at the design table.
- **Standby redundancy** - A fault-tolerance approach where a backup unit stays idle until the primary fails, then takes over. It saves the operating cost of running the backup but introduces a switchover gap that can itself fail.
- **Sustainability** - Designing systems that account for their environmental impact across their whole lifecycle, from raw materials and manufacturing through operation to disposal, treating environmental cost as a real cost rather than an externality.
- **SysML** - The Systems Modeling Language, a standardized notation for representing systems as connected diagrams of structure, behavior, and requirements, used to build the linked models of model-based systems engineering.
- **System lifecycle** - The full span of a system's life, from the first concept through development, production, operation, and retirement. Early decisions commit most of the lifecycle cost while the system still looks like drawings.
- **Systems engineering** - The practice of treating a system as a single thing to be designed even though it's made of many parts built by many people, covering requirements, decomposition, interfaces, verification, and the cost-schedule-performance trade-offs across its life.
- **Systems thinking** - The ability to see a problem as a web of parts that affect each other, often in loops, rather than as a straight chain of cause and effect. It catches the errors that linear reasoning produces in complex systems.
- **Tacit knowledge** - The things stakeholders know so well they've forgotten they know them, like the unwritten rules and exceptions an experienced worker follows. It's exactly what a system must honor and exactly what people forget to mention.
- **Technical debt** - The accumulated cost of choosing expedient shortcuts over sound solutions, carrying interest because the shortcuts make future work harder. Taken on deliberately it can be sound; left unmanaged it can strangle a project.
- **Technical performance measure (TPM)** - A specific technical parameter tracked over a project's life against where it should be, giving early warning when the system is drifting away from its requirements while correction is still cheap. Weight is the classic aerospace example.

- **Test matrix** - A structured grid mapping test cases against the requirements or conditions they cover, used to expose gaps in coverage that would otherwise hide behind a vague feeling of having tested thoroughly.
- **Top event** - The specific catastrophic event a fault tree is built to prevent, sitting at the top of the tree with all its possible causes branching downward beneath it.
- **Trade study** - A structured analysis comparing alternatives that differ in cost, schedule, performance, and risk, against weighted criteria, so a hard choice is made deliberately and defensibly. Its integrity depends on setting criteria honestly rather than to reach a predetermined answer.
- **Validation** - Confirming that you built the right system, that it actually meets the real operational need. It's the only safeguard against requirements that were wrong, because verification trusts the requirements completely.
- **Verifiable requirement** - A requirement you can determine, by an objective method, whether it's met. If you can't describe how you'd prove it's satisfied, the requirement is broken however reasonable it sounds.
- **Verification** - Confirming that you built the system right, that it meets its specifications. It says nothing about whether the specifications were correct, which is the separate job of validation.
- **Voting redundancy** - A fault-tolerance approach where several units perform the same computation and a voter takes the majority answer, so the system tolerates not only a unit failing but a unit producing a confident wrong result.
- **Work breakdown structure (WBS)** - A hierarchical decomposition of all the work in a project into progressively smaller pieces, complete and non-overlapping, forming the backbone that the schedule, budget, assignments, and tracking all hang on.
- **Work package** - The smallest piece in a work breakdown structure: a chunk of work small enough to assign a single owner, estimate, schedule, and track. The level where planning and tracking actually happen.

Works Cited & Recommended Readings

Billah, K. Y., & Scanlan, R. H. (1991). Resonance, Tacoma Narrows bridge failure, and undergraduate physics textbooks. *American Journal of Physics, 59*(2), 118–124.

Blanchard, B. S., & Fabrycky, W. J. (2011). *Systems engineering and analysis* (5th ed.). Prentice Hall.

Boehm, B. W. (1981). *Software engineering economics*. Prentice Hall.

Brooks, F. P. (1995). *The mythical man-month: Essays on software engineering* (Anniversary ed.). Addison-Wesley.

Columbia Accident Investigation Board. (2003). *Columbia Accident Investigation Board report* (Vol. 1). National Aeronautics and Space Administration.

Conway, M. E. (1968). How do committees invent? *Datamation, 14*(4), 28–31.

Dastin, J. (2018, October 10). Amazon scraps secret AI recruiting tool that showed bias against women. *Reuters*.

Eykholt, K., Evtimov, I., Fernandes, E., Li, B., Rahmati, A., Xiao, C., Prakash, A., Kohno, T., & Song, D. (2018). Robust physical-world attacks on deep learning visual classification. *Proceedings of the IEEE Conference on Computer Vision and Pattern Recognition (CVPR)*, 1625–1634.

Feynman, R. P. (1986). Personal observations on the reliability of the shuttle. In *Report of the Presidential Commission on the Space Shuttle Challenger Accident* (Vol. 2, Appendix F). U.S. Government Printing Office.

Friedenthal, S., Moore, A., & Steiner, R. (2014). *A practical guide to SysML: The systems modeling language* (3rd ed.). Morgan Kaufmann.

Goodfellow, I. J., Shlens, J., & Szegedy, C. (2015). Explaining and harnessing adversarial examples. *International Conference on Learning Representations (ICLR)*.

Greenberg, A. (2015, July 21). Hackers remotely kill a Jeep on the highway—with me in it. *Wired*.

International Council on Systems Engineering. (2015). *Systems engineering handbook: A guide for system life cycle processes and activities* (4th ed.). Wiley.

Kemeny, J. G. (1979). *Report of the President's Commission on the accident at Three Mile Island*. U.S. Government Printing Office.

Leveson, N. G. (2011). *Engineering a safer world: Systems thinking applied to safety*. MIT Press.

Leveson, N. G., & Turner, C. S. (1993). An investigation of the Therac-25 accidents. *Computer, 26*(7), 18–41.

Levinson, M. (2006). *The box: How the shipping container made the world smaller and the world economy bigger*. Princeton University Press.

Lions, J. L. (1996). *Ariane 5 flight 501 failure: Report by the inquiry board*. European Space Agency.

Meadows, D. H. (2008). *Thinking in systems: A primer*. Chelsea Green.

National Aeronautics and Space Administration. (2016). *NASA systems engineering handbook* (Rev. 2, NASA/SP-2016-6105). NASA.

National Transportation Safety Board. (2019). *Collision between vehicle controlled by developmental automated driving system and pedestrian, Tempe, Arizona, March 18, 2018* (NTSB/HAR-19/03). NTSB.

Norman, D. A. (2013). *The design of everyday things* (Rev. ed.). Basic Books.

Perrow, C. (1999). *Normal accidents: Living with high-risk technologies* (Updated ed.). Princeton University Press.

U.S.–Canada Power System Outage Task Force. (2004). *Final report on the August 14, 2003 blackout in the United States and Canada*. U.S. Department of Energy & Natural Resources Canada.

U.S. Securities and Exchange Commission. (2013). *In the matter of Knight Capital Americas LLC* (Release No. 70694). SEC.

Vaughan, D. (1996). *The Challenger launch decision: Risky technology, culture, and deviance at NASA*. University of Chicago Press.

Vesely, W. E., Goldberg, F. F., Roberts, N. H., & Haasl, D. F. (1981). *Fault tree handbook* (NUREG-0492). U.S. Nuclear Regulatory Commission.

Zetter, K. (2014). *Countdown to Zero Day: Stuxnet and the launch of the world's first digital weapon*. Crown.

References & Recommended Readings

Kossiakoff, A., Sweet, W. N., Seymour, S. J., & Biemer, S. M. (2011). *Systems engineering principles and practice* (2nd ed.). Wiley.

A well-organized textbook that walks through the full discipline in more technical depth than this book attempts, useful if you want the formal methods laid out step by step. Good for readers who want a more conventional engineering textbook treatment of lifecycle processes, requirements, architecture, integration, and verification.

Petroski, H. (1985). *To engineer is human: The role of failure in successful design.* St. Martin's Press.

A readable, engaging argument that failure is how engineering learns, full of concrete cases that bring the lessons of this book to life.

Reason, J. (1990). *Human error.* Cambridge University Press.

The source of the influential "Swiss cheese" model of how accidents happen, essential if the human-factors material was the part that grabbed you.

Robison, P. (2021). *Flying blind: The 737 MAX tragedy and the fall of Boeing.* Doubleday.

A gripping narrative account of how organizational and management failures, not just engineering ones, produced a disaster, and a sobering case study in the themes of Chapters 5 and 7.

Senge, P. M. (2006). *The fifth discipline: The art and practice of the learning organization* (Rev. ed.). Doubleday.

A classic that takes systems thinking out of engineering and into how organizations behave, a natural next step if Chapter 1 changed how you see problems.

Sterman, J. D. (2000). *Business dynamics: Systems thinking and modeling for a complex world.* McGraw-Hill.

A thorough, technical treatment of feedback loops, delays, and system behavior. For readers who want to actually model the dynamics rather than just understand them conceptually. Widely regarded as one of the foundational texts in system dynamics, it bridges theory and practical application through extensive real-world examples and simulation-based analysis.

www.ingramcontent.com/pod-product-compliance
Lightning Source LLC
LaVergne TN
LVHW021736071126
851857LV00040B/1456
9798900421056